Heat Pumps
& Houses

Murray Armor

PRISM PRESS

First Published in 1981 by

PRISM PRESS
Stable Court
Chalmington
Dorchester, Dorset DT2 0HB.

ISBN O 907061 19 2 Paperback
ISBN O 907061 20 6 Hardback

Typeset by Ryton Typing Service, Worksop.
Cover by Rich Designs, Worksop.

Printed in Great Britain by
Billing & Sons Limited
Guildford, London, Oxford, Worcester.

CONTENTS

INTRODUCTION

Heat pumps are news; in the last year almost every magazine and Sunday newspaper has discovered them. We read they offer a new cheap way of heating our homes, and a way to escape from spiralling fuel costs. They are becoming a standard feature of 'House of the Future' exhibitions, and manufacturers take a lead from this and enthusiastically plug them as being new, trendy and revolutionary.

All this rather ignores that the man in the street who is proposing to make a major investment in a new central heating system is usually very cautious in making his choice, and is disinclined to spend his money on something which he feels may not yet be fully proved. On the other hand, fuel prices have increased so far in front of wages that any possibility of making a major saving in energy costs demands consideration. The man in the street knows this too.

It is claimed that the savings to be made can justify pulling out perfectly good oil fired central heating appliances and replacing them with heat pumps, or at least installing a heat pump to work together with an existing boiler or warm air unit. We are told that a heat pump installation in a new house is even more economical. The theory seems irresistible. What of the practice?

This book looks at heat pumps from the purchasers view point, looking at what he can expect to find for sale, what it will do, what it will cost to run, and what the snags may be. It is written for the householder with a fuel bill problem, and is about homes that are heated with heat pumps, and not about the technicalities of the machines themselves. A large part of the book is taken up with case histories, which are all of ordinary domestic installations and not research projects. It is not about experiments, but about machines which are on the market now.

If *HEAT PUMPS AND HOUSES* persuades the reader that he should look further into a better way of heating his own home, I hope it will also have persuaded him that he will need expert professional advice. The book should help him to understand and evaluate the advice, and will suggest some questions for him to ask his advisor. It is not suggested that anyone should consider a heating installation of this sort without the help of a good heating engineer.

An apology over the use of °F. and °C. A great deal of time has been spent debating whether temperatures quoted should be expressed in the Farenheit or Centigrade scales. The argument that prevailed is that those familiar with °C. will usually be able to convert to °F, but not vice versa. The use of the Farenheit scale will infuriate the Scientists or Engineers who may read this. I apologise. The same is true of possible abuse of the strict definitions of kW and kWh. Again I apologise. One cannot be all things to all men; I hope that I have been of use to the layman.

My thanks are due to officials of the Electricity Council and various Electricity Boards who have given help and advice in collecting material for this book, and to the many manufacturers who have made information available. Above all, I am indebted to the people who have let me feature their own heating installations in the case histories. Fortunately those who are connected with a fast moving technology tend to be both enthusiasts, and to be generous in sharing their enthusiasm. Unlikely as it seems, heat pumps are fun to a lot of people. While I have been writing this book (and also installing my own heat pump) I have found it fun as well. I hope that the reader will.

CHAPTER ONE

HEAT PUMPS AND HOMES

Heat pumps are popularly presented as being new and revolutionary inventions which offer a dramatic saving in domestic heating costs. Like most popular views of a new technology, this can be misleading. Heat pumps are neither new nor revolutionary, and they are not the automatic answer to every problem with heating bills. What they do offer is an opportunity to heat most homes more cheaply than by using some fuels, using a technology that has been with us for nearly a century, but which has only recently become relevant to domestic heating due to the dramatic rise in the cost of all fuels.

A heat pump is a machine that collects heat from one place and moves it to another place. It does not generate any heat itself. This is a particularly difficult concept to understand on first meeting it, and how this is done is even more difficult to understand. Consideration of the latter — how the machine works — is firmly relegated to appendix one, and it is strictly for the technically minded. However, knowing what a heat pump does is essential to any serious consideration of using one of these machines, and it is fortunate that we have a familiar example in almost every kitchen. A domestic refrigerator is a heat pump which moves heat out of the cabinet and dumps it to waste via the warm coils hidden somewhere at the back. A deep freeze also has a heat pump, pumping heat out of the freezer chest and using a fan to blow it to waste as warm air whenever the pump is working. If your local supermarket is a modern building it is likely that the warm air which is blown at you as soon as you enter the door is the heat pumped out of the refrigerated display cabinets. The technology that makes this possible is refrigeration engineering, and ever since the 19th century it has only been concerned with keeping things cool. For all of this time it has also been possible to reverse refrigeration pumps and pump heat into a cabinet or into a building. Refrigeration pumps are heat pumps, and they can collect heat from the air, or from a river, and pump it anywhere it is required. The air or the river may not feel warm, but a heat pump can extract plenty of heat from them leaving them colder still. It can pump this heat into a house, and release it to raise the temperature in the house to any domestic level that is required. The cost of doing this, taking into account the cost of the equipment, has until recently been much more than the cost of providing the same heat by burning oil, gas or coal, or by using electricity, so there was no point in it.

The world rise in fuel costs since the mid-1970s has made the cost of burning fuel in a heating appliance far more important than the cost of the appliance itself, and it is this which has now made heat pumps very interesting alternatives to fuel burning appliances.

In 1964 the author's oil fired central heating installation, serving nineteen radiators, cost £500 to install and £70 per annum in oil fuel. A maintenance contract, including the insurance of motors and pumps, cost under £10 per annum. Running costs were 16% of capital costs.

In 1980 the same installation would cost well over £1,700 to install and the actual fuel costs and maintenance contract in the year came to over £850. Running costs were 50% of capital costs.

This shift in emphasis from capital costs to running costs is critical to any consideration of a new heating system, or to the up-dating of an existing heating system. Running costs dominate all, and have now reached a level where the high capital costs of a heat pump can be covered by the savings made in running it over relatively short periods. Everything depends on the fuel with which a comparison is to be made, the level of heating required, and the house to be heated. As will be discussed later, a heat pump today does not offer any advantage over a gas fired central heating system, and very little cost advantage over a solid-fuel installation. On the other hand, the expense of a new heat pump replacing an oil boiler, or replacing some forms of electric heating, will be covered by the savings in running costs in a relatively short while. Of course, heat pumps do use electricity, but they are able to pump up to three times as much heat into a building with one unit of electricity compared with the heat that would be generated by using the same unit of electricity in an electric fire or storage heater. The steady increase in world fuel prices can only make it progressively cheaper to use electricity to move heat about than to use electricity in a simple heater.

This book looks at the use of heat pumps to provide central heating for houses as a simple alternative to using heating appliances that burn fossil fuels — oil, gas or coal. It is concerned with every aspect of using equipment that is commercially available today. The emphasis is on the house, and on the people living in it, and not on the gadget itself. Heat pump technology and a detailed description of how the things work are from now on firmly relegated to an appendix, for if heat pumps are to become the basis of our heating systems of the future there is no reason why householders should have any better understanding of how they work than they have of how their television sets work. On the other hand, the householder knows very well how to use his television set, how to tune the automatic controls, and has firm opinions about the best position for the aerial and the relative merits of different makes of set. If the heat pump is to realise its potential as a domestic heating appliance, we will have to learn to understand its characteristics as well as we understand those of the television set, even though its workings may be as much of a mystery.

A heating installation in a home is required to have three features: it has to provide the heat that it is required to provide, it has to do so in a way that is convenient to the householder, and it is excepted to do so at an acceptable cost. To understand what the heat pump can do to meet these needs it is necessary to look at the needs themselves in rather more detail.

Firstly, definitions of central heating need looking at carefully. For thirty years the fuel companies and heating appliance manufacturers have directed their advertising to persuading us that a properly heated house is one in which the occupants can relax wearing the lightest of clothes with the coldest of weather outside. This follows the fashion across the Atlantic, where a combination of cheap fuel, free steam heat in apartment blocks, and a Hollywood portrayal of all America enjoying a Californian climate have led to indoor temperatures which we criticised as unhealthy in the '50s, and had generally adopted by the '70s. Between the wars central heating in this country was generally taken to imply temperatures 55°F. to 60°F. By the '60s the term "central heating" had been defined to refer to a house with equipment capable of 65°F. An installation giving lower temperatures was to be described as "back-ground heating".

By 1981 the required standards for central heating are 70°F. in the lounge and 60°F. in bedrooms. Back-ground heating is rarely mentioned.

Although a disproportionate amount of energy is used in heating a building through the last 10°F. to reach the designed temperature, no-one has yet dared to suggest that these standard temperatures

A period house and its heat pump. This home had no central heating at all until the heat pump was installed. It is a combined unit, with all the machinery in the casing outside. It collects heat from the air and transfers it to water which circulates in conventional radiators.

Three heat pumps working together to heat a large country house in Nottinghamshire. They are split units, with the compressors in a stable block and their absorber units outside. In the summer this installation heats a swimming pool.

are too high, even though they were generally scoffed at as unhealthy, if not degenerate, a generation ago. The short lived "Save It" Campaign of 1979 advocated turning down thermostats, but stopped short at suggesting the level at which they should be set. It is possible to argue that this is part of the national refusal to contemplate any reduction of living standards in the face of our changing economic circumstances. Be this as it may, the energy required to heat the average new home drops from 25,000 KWh a year to meet the full central heating standard to 17,500 KWh a year if it is only to be heated to a 60°F. lounge/50°F. bedroom level. The saving is 30%.

In the face of soaring fuel costs many people have already made their own decision in this, and have turned down the thermostat and bought heavier underwear. Some are on low incomes and are literally unable to afford high fuel bills, but many more are those in large houses with four-figure fuel bills who found that they had to choose between 60°F. in their homes and an extra holiday break, or 70°F. and no holiday. Many have chosen the former. Anyone considering a new heating system of any sort must consider what their approach is to this, and whether they intend to continue with the shirt-sleeves life style of the central heating advertisements, or whether they will settle for keeping their jackets on and having significantly lower heating bills. Incidentally, a little remarked difference between state and private education is that private schools, justifying parsimony by reference to the spartan ideal, keep their premises at least 10°F. lower than local authority schools where the rate-payer foots the fuel bills. Children living at the lower temperature do not seem to suffer any ill effects.

The significance of all this in a book about heat pumps is that it is particularly important that the unit installed should suit the heat demand of the house if both capital expense and running costs are to be kept to a minimum. If you are considering buying a heat pump costing considerably more than a conventional boiler or air heater because you want the benefit of the huge saving in running costs, there is no point in getting a larger machine than you need. On the other hand, if you want 70°F. and shirt sleeve living, then a heat pump installation of the right size will certainly meet your requirements and will make even larger savings over the cost of burning oil. The right size of unit is all important, and this can only be assessed by a Heating Engineer who can calculate the heat required to warm your house to the temperature *you* want.

The next consideration is the convenience of your heating system. There are four basic ways in which a room can be centrally heated — using radiators, warm air, underfloor heating, or ceiling heating. (Night storage radiators are not now classed as central heating.)

A radiator system uses hot water circulating through pipes from the heating appliance to the radiators, but can also supply this hot water to warm air fan heaters. For this reason the phrase "radiator system" is confusing, and we shall use the trade term "wet system" from now on.

Warm air systems (as opposed to warm air heaters in wet systems) use air which has been heated in a central unit, and which is blown through ducts to outlets in all the rooms, where it usually enters through grilles in the floor or walls.

Underfloor heating involves heating up the whole floor slab using cheap off-peak electricity, and letting this heat diffuse out into the house all day without a great deal of control over it. Although the heating is usually provided by wires embedded in the concrete, it is possible to use a heat pump to warm a floor slab that has pipes embedded in it.

Ceiling heating relies on radiant panels set in a specially installed ceiling. It is a relatively new technology, and has only a very small share of the market in spite of being heavily promoted by The Electricity Boards.

A heat pump can be used to provide heat for both wet and warm-air systems, either as initial equipment in a new home or to replace or augment existing boilers or warm air heaters. It can be used for an underfloor heating system, but this, together with ceiling heating, steam heating and other unusual systems are outside the scope of this book. So are full air conditioning systems, which although based on heat pumps are unlikely ever to come into general use in British homes.

The relative merits of wet and warm-air systems are frequently debated, and decisions between them are a personal judgement. A chart giving the pros and cons is given overleaf. Hot air heating had a considerable vogue in the Sixties, but has since lost much of its market share in new housing. It does, however, have one interesting advantage over wet systems for use with a heat pump: as soon as the central heating is turned on there is an almost immediate response. This is important to many people who use their central heating system to provide heat only where and when they want it, and who turn their system on like an electric fire when they feel chilly. A warm-air system will deliver warm air almost immediately, and although this may take an hour or more to bring the whole house up to temperature, it is detected straight away as one moves about the building, and this gives assurance that the system is working. To a lesser extent a wet system that is connected to a boiler with the controls at a maximum setting gives the same psychological effect, for the radiators will be too hot to touch long before the whole house is warmed through. On the other hand, a heat pump connected to a wet system usually heats larger radiators to a rather lower temperature than an oil or gas boiler, and so does not give the same assurance that a localised heat source does. The radiators served by a heat pump will feel warm while they warm the whole house; the radiators served by an oil boiler may feel hot while they do the same thing; a warm-air system will waft out reassuringly warm air however it is heated. Human attitudes are often illogical, and for may people it is as important to feel that they are getting warm as it is to actually get warm. While collecting data for this book one person who has happily changed from oil to a heat pump expressed it by saying "it gets the house just as warm, but the radiators don't get as hot, so you don't know its getting warm". If radiators that are too hot to touch are an essential part of your understanding of what a central heating system should be, you will have to do some adjusting if you want a heat pump. On the other hand, if you have a hot air system you will not notice the difference.

The ideal use of the heat pump in a domestic heating system is to provide a steady temperature level throughout as long a heating period as possible, and they are not at their best if used as a heat source that is expected to provide warmth at the flick of a switch. They are well suited to elaborate control systems, and they will be at their most effective, and most economical where their controls are set up to provide the required temperature and the system is then left to get on with its job.

COMPARISON OF WET AND WARM AIR HEATING SYSTEMS

WET	WARM AIR
Can be installed in existing houses and can be altered easily.	Involve big air ducts built into the structure — must be allowed for when the house is built.
Radiators take up wall space and tend to inhibit furniture arrangements.	Outlet grilles do not take as much space and permit more flexible furniture arrangements.
Cannot be the basis of other environmental systems.	Can incorporate air filters to cut down dust, and can have automatic humidifiers. Can be the basis of a full air conditioning system.
Slow response when turned on.	Fast response when turned on.
Universally acceptable.	Some asthmatics and others find warm air systems objectionable.
Thermostatic valves can be fitted to individual radiators.	Individual outlets can be closed but automatic controls for individual rooms not generally available.
Noise problems unlikely, and relatively easily cured.	Any transmitted noise is very difficult to cure.
No prejudice against wet systems.	Prejudice against warm air heating may inhibit sale of property.

The convenience of a heating system also requires that it shall be unobtrusive, clean, labour-saving and requiring little maintenance. Compared with its fossil fuel competitors the heat pump has a clear advantage in three of these four considerations. It is not only clean, but is non-polluting, with no smoke or flue gases discharging into the atmosphere. It is as quiet as an oil boiler, and does not require fuel deliveries or the fuel storage arrangements involved with oil or solid fuel. It requires less maintenance than any fuel burning appliance. However, the largest category of heat pumps, those that take heat from the air, are certainly not unobtrusive. Whatever the design of the particular unit, it must incorporate several square feet of heat exchanger coils which have to be put somewhere where the air flow around them is uninterrupted. They are normally cased in a unit of at least the size of a coal bunker, and only marginally more attractive. Because of the need not to restrict the air flow, they cannot be disguised by hiding them in a shrubbery. This can be best considered by studying the various photographs. Further to this, if the heat pump is to have a heat store, there is another 100 cubic feet of tank involved, although this does not require to stand in the open and can be hidden in a cellar or out-building.

Turning now to our third requirement for heating installations, that of cost, it is again useful to first take a distant view.

The shift in emphasis from capital cost to the running costs is critical to any consideration of a new heating system, or to the up-dating of an existing heating system. Running costs are all-important, and to understand them one has to look at the "useful kilowatt", for useful kilowatts are what the householder buys whatever fuel is purchased.

Until the advent of metrication, heat was measured in British Thermal Units or B.T.Us. This was a very easily recognised term for describing units of heat, and so was abandoned and replaced with the incomprehensible kilowatt, previously used only to measure electrical energy. However, energy is heat, and in this book the term is used to measure a specific amount of heat. Quantities of electricity will be measured in units, as this is the term used on electricity bills. "Useful kilowatts" (or kW) are used to measure the quantity of heat that is actually obtained from burning a particular fuel. Kilowatt hours or kWh are the units used to express the number of kilowatts used in one hour. It is easy to confuse kW with kWh, and for our purposes the difference between them is academic.

A 3 bar electric fire burning in a room will generate 3 kW of heat, all of which are useful kW which go into the room and warm it up. A gallon of oil burning in an oil fired central heating system has an energy content of 48 kW of heat, but only 31 kW of this will leave the radiators and warm the rooms, the rest going up the boiler flue or being lost in other ways. Only the 31 kW are useful kilowatts.

Most boilers and other heating appliances have fairly standard efficiencies and give the same set output of useful kilowatts from standard quantities of fuels. From this we can arrive at costs per useful kWh for different fuels, and this is easily expressed in pence per kWh. This leads to the following comparisons at spring 1981 prices.

OIL
Oil prices are used as an economic regulator by Governments, and duty on heating oil is a source of revenue. There seems little chance of oil prices ever moving down. Apart from this, oil is a diminishing resource and is more valuable as raw material for the chemical industry than as fuel, which must keep the prices up in the future. Oil already costs 2.5 pence per useful kWh, is the most expensive fossil fuel, and unlikely to cease to be so.

GAS
Gas is the most economical fuel for central heating available today, costing 1.1 pence per useful kWh. In the last analysis the price is controlled by politicians, but gas is so widely used as a fuel by so large a cross section of the population that it is unlikely that the price will rise to the level of oil prices while gas still flows from the North Sea although the Government has announced its intention of raising gas prices by 10% p.a. in real terms. Allowing for inflation this will increase gas heating costs by 25% p.a.

SOLID FUEL
The coal industry has a long term future, but it is unlikely that solid fuels will ever get any cheaper than they are at present. Different fuels have different costs, but Sunbrite, a typical domestic boiler fuel, costs 1.6 pence per useful kWh. Using solid fuels inevitably involves handling both fuel and ashes, which is becoming progressively less acceptable in a labour saving world.

ELECTRICITY USED FOR DIRECT HEATING
Full tariff or day-rate electricity is the most expensive form of energy, costing an average of 4.5 pence per useful kWh. Economy 7

tariff electricity is much cheaper at an average of 1.82 pence per useful kWh but is only available from midnight to 7.00 a.m. If energy purchased at the ecomony rate can be stored as heat, as in a night storage heater, it can be used during the day with a dramatic reduction in heating costs. The figure will depend on the heat losses involved in the process, and with a storage/retrieval system working at 80% efficiency the cost becomes about 2.3 pence per useful kilowatt.

The fifteen different Electricity Boards all have slightly different prices: the figures quoted are the averages. The actual prices charged by the different boards are set out in Appendix 2.

This neat inconspicuous Myson unit provides 7.5 kW of heat for a consumption of 2.8 kW of electricity, and is featured in Case History Four.

ELECTRICITY USED TO OPERATE A HEAT PUMP

This cost is really quite different from the others, in that it is not a measure of electricity directly converted to heat, but of electricity used to collect heat from elsewhere and to pump it into the building. On full tariff this is 1.8p per useful kWh, and on the night rate of an Economy 7 tariff it is .7p. Night rate electricity is only available for 7½ hours in every twenty-four, but it is usual to run heat pumps on this tariff to give a lower cost for heat used at night. Cheap rate electricity can also be used to put cheap heat into a heat store at night, so that it can be used during the day.

This cost depends on the ability of the heat pump to use one unit (or kWh) of electricity to pump between 2 and 3 kWh of useful heat from the heat source into the house.

We can now translate these rather abstract figures into estimates of likely fuel bills. The average 1200 sq.ft. house — typically the small four bedroomed house with two reception rooms — has an annual heat and hot water requirement of 25,000 kWh if it has up-to-date standards of insulation. This same figure will cover the heating requirements of a much larger house of up to 2000 sq.ft. if it has an unusually high level of insulation. The fuel bills to be expected with the different fuels are thus:—

DAY RATE ELECTRICITY	
(ie. electric fires or convector heaters)	£1100 p.a.
ECONOMY RATE ELECTRICITY	
(ie. night storage heaters or underfloor heating)	£422 p.a.
OIL	£625 p.a.
SOLID FUEL (as Sunbrite)	£400 p.a.
GAS	£275 p.a.
HEAT PUMP, Full tariff	£450 p.a.
HEAT PUMP, Economy tariff	£175 p.a.
HEAT PUMP	
⅓ Economy tariff, ⅔ Full tariff, (usual situation)	£350 p.a.

These are average figures in the Spring of 1981, based on average fuel costs. They are generally borne out by the actual costs in the case histories that follow, but there are other considerations, and we must now move on to discuss them.

CHAPTER TWO

HEAT PUMPS — THE ECONOMICS

So far we have seen that a heat pump used with full tariff electricity offers running costs significantly below those for oil fuel or full tariff electricity, that these costs are on par with solid fuel or off-peak electricity costs, and are higher than those for gas. If the heat pump is used on an Economy 7 tariff to heat up a heat store of some sort at night, its running costs are far lower than any other sort of central heating system. We now have to consider capital costs.

There are two sets of circumstances in which a heat pump may be installed in a home. Either the unit is to replace or work in conjunction with an existing appliance which is in working order but which is costing too much to run, or else it is being put into a new house, or one where an existing appliance is being replaced because it has come to the end of its life. In the first case the whole of the cost of the equipment and installation has to be set against the expected savings, in the second a separate set of calculations has to be made. These two alternatives can be set out as follows:—

REPLACING EXISTING APPLIANCE

> Cost of heat pump and installation
> Divided by anticipated saving per annum
> Equals pay-back period for heat pump installation.

INSTALLED IN NEW HOUSE OR REPLACING APPLIANCE WHICH HAD TO BE REPLACED ANYWAY.

> Cost of heat pump and installation
> Less cost and installation of cheaper alternative
> Divided by saving in running costs over cheaper alternative.
> Equals pay-back period for heat pump installation

The actual way in which these calculations are made is best examined by looking at real situations. Case History 3 deals with a warm air heating system where the heat pump installation was made at a cost of £1,256. The saving in 1980 was at a rate of £152 per annum, which gives a pay-back period of 8 years. If we anticipate a 10% p.a. inflation in fuel prices, the pay-back period becomes 6 years. The house-holder, who is an architect, considers this very good value.

DECEMBER 1979 DATE.	HOUSE OUTSIDE T°F 8 AM	HOUSE OUTSIDE 12.30	Roof TEMP. 8 AM	Roof TEMP 12.30.	UNITS ALL USAGE	A 6-10pm	B 4-8am	C 8-6pm	NOTES.
									Room stat set 65°F
1	49°	50°			49	13	.16	20	
2	50°	51°			48	13	15	20	Pump Stat 69°C
3	47°	50°			45	12	15	18	
4	48°	52°			49	10	18	21	
5	48°	50°			49	14	16	19	
6	48	52			56	12	20	24	
7	50°	51°			48	10	19	19	
8	48	50.			53	13	16	24	
9	48°	49°			54	12	17	25	
10	44	48			62	11	19	32	
11					62	10	20	32	Central Heat cut out at
12	42°				75	13	21	41	9.30 am.
13	46°				55	11	20	24	Central heating off at 5.30
14	42°	46°	36°	46°	63	12	20	31	2lbs Gas put in .
15	38°	-	38°		67	10	21	36	CH off 5.15.
16			36°		66	11	22	33	
17					69	11	22	36.
18				40.	69	12	21	36	
19	40°				74	11	21	32	
20	36°				76	14	20	42	Meter changed to 8 am
21	36°				77	16	25	36.	
22	30°				77	15	25	37	
23	38°		30°		82	14	26	42	
24									
25									
26					away.				
27									
28					504.	-	-	-	
29.	40°		36.		76.	14	24	38	
30	40°		32		70	14	24	32	
31.	38°		30°		70	14	24	34	
			TOTAL UNITS=1935.			(320	660	955)	

Domestic usage 26 × 9 = 234 =. 1701 heat pump. ÷ (18 × 31) = 558) = 3.048 KW/Hour so the pump was running about 80% of the month.

Many of those who have a heat pump installed in a home that was costing too much to heat keep records of the performance of their new installation for a while. This is a page of a log kept of electricity consumption and temperature for the first months performance of a heat pump installed in a bungalow in Wales. 1935 units were used. The cost was £64 as some were on the cheap rate. This was in December 1979.

Case History 7 details a wet system where an oil boiler was pulled out and replaced with a heat pump which was supplied and installed for under £1000. Air ducting installed by the occupier cost another £180. The annual saving is £208 and the pay-back period is under six years without any consideration of inflation.

The application of these figures to ones own circumstances depends on two things. The first is a professional appraisal of the cost and probable performance of a heat pump in ones own home, and the other is ones own view of the value of an investment in energy saving taking into account the likely pay-back period. Conventional financial wisdom is that a pay-back period of 8 years is a sensible investment for almost anyone. If you do not have your own firm opinion on this, it is worthwhile discussing it with your bank manager or professional advisor.

All of this has assumed that the heat pump is going to provide all the central heating required in the home. This is not always the case. It is frequently sensible to retain an existing oil-fired boiler or other heating system, and to install a heat pump to work with it, the heat pump providing a base load of heat with the old boiler starting up on automatic controls when a cold house has to be heated up from scratch, or in exceptionally cold weather. This is described by manufacturers as a bivalent system, and it is a very practical proposition indeed where space permits the two separate installations. The cost equations are the same; capital cost divided by saving equals pay-back period, and these installations are often particularly worthwhile.

Use of a heat store to take full advantage of the Economy 7 tariff offers the most convincing savings of all, but raises problems of where to put the store itself. The cost of this can be very little if a suitable tank can be obtained cheaply, and the savings can be very significant. The author's heat pump has a 500 gallon hot water heat store. At midnight the temperature of this water is 110°F. Economy 7 tariff electricity then heats it to 142°F. by dawn, at an average expenditure of 50p on 30 units of cheap rate electricity. During the day this hot water is pumped around the radiators for at least five hours, and often longer, before its temperature drops to 110°F. again. The heat pump then starts up automatically and heats the radiators directly with the heat store disconnected. At midnight, when the cheap rate period starts, the heat store is automatically connected into the system and the heat pump warms it to 142°F. by the end of the cheap rate period. There are 19 radiators. A typical

A heat pump with a heat store tucked away in the cellar of a seventeenth century house, where the lady of the house keeps a daily record of the electricity consumption on the back of the cellar door! There are two separate water tanks linked together to hold the heat collected at night using Economy 7 tariff electricity: they can be seen wrapped in aluminium foil insulation.

week's consumption is 226 cheap rate units and 187 full rate units, costing £12.05. If all were full rate units the cost would be £18.17. The storage arrangement alone saves £6 per week.

It is always interesting to speculate whether a heat pump heating system adds significantly to the value of a house on resale. The answer is probably not, but that it will give a house an edge over another with less economical heating at the same price. The analogy is with gas heating: estate agents advise that gas does not add to the value of a property, but that it will help it to sell more readily than a similar property heated with oil.

A further financial consideration is that V.A.T. is charged on all heating equipment, including heat pumps, which are purchased directly by private individuals, but not on complete installations undertaken by installers. As a result of this there may be little advantage in buying equipment yourself and hoping to arrange for the installation to be carried out an a labour-only basis.

Virtually all heat pump manufacturers will offer financial facilities themselves, or will put you in touch with someone who will.

A heat pump installation in a new home will certainly be covered by your mortgage.

CHAPTER THREE

INSULATION

It is taken for granted that anyone buying a book on heat pumps is well aware of the need for modern standards of insulation to cut fuel costs for any heating system, and there is no need to labour this point except to point out that the more sophisticated the installation, the more important this is. For the average house a heat pump is the most sophisticated installation available, and the insulation must match it, at least as far as it is cost effective. However, deciding what is going to be cost effective is often difficult. All houses have differing existing levels of insulation, and different insulating techniques will give significant savings or a negligable saving depending on the characteristics of the individual building.

In addition to this, expenditure on insulation can be regarded as either a direct investment, with a return expected immediately, or as an indirect investment which is not immediately cost effective but which helps to make the home more comfortable, and will become cost effective as inflation forces up heating costs. It is essential to consider which group you are in, and also to note that there can be another group: those who wish to insulate a dwelling to a standard which is not cost effective in relation to direct heating costs, but who hope to get their heat loss figure down to a level where they can heat the place with one heat pump instead of two, making a huge capital saving. This last group needs specialist advice and should be able to get it from a heat pump installer. For the rest, the rules are as follows:

If you are having a new house built on your own land, discuss insulation levels and costs with your architect at a very early stage. If he knows you are thinking of a heat pump he has probably raised the subject with you anyway. Ask him to look into the use of cavity slab insulation with you: this involves building sheets of insulating material into the walls while retaining the traditional cavity, giving the best of both worlds.

If you are buying a new house that is already planned, and are considering asking for a heat pump to be installed, ask for a written statement of the designed insulation values, and for the details of additional insulation available. Discuss this with the heat pump installer, and order the additional insulation from the builder. In this way you will get it on your mortgage, and will be protected by the builder's N.H.B.C. warranty if anything goes wrong.

If you have an existing house keep in mind that the batting order in terms of being cost effective, is —

* Sophisticated draught seals to windows and doors. There is no longer any need for this to involve unattractive looking strips of self-adhesive foam: ask to look at all the options at a good building centre, or write for a catalogue to Schlegel (U.K.) Ltd., at Henlow Industrial Estate, Henlow Camp, Bedfordshire SG16 6DS. Draught-proofing is easily the most cost effective and energy saving ploy.

* Blanking off unused chimneys. It is important to have this done properly and to get advice on maintaining the necessary minimum ventilation in the flue.

* Up to 8'' of roof insulation.

* Porches at the front and back doors if these are aesthetically and functionally desirable.

* Double glazing, although this is rarely immediately cost effective and the pay-back period may be very long. On the other hand, a double glazed house is much nicer to live in if you lead a sedentary life and like to sit near your windows. Choose your double glazing from personal recommendation and from looking at friends' homes, and not from trade literature. If you have a house with badly insulated walls, insulating them may deserve priority over double glazing.

* Wall insulation. A very complicated matter, as follows.

Houses built from mid-1981 will have to have new and very high insulation standards, and no additional insulation to the walls is likely to be cost effective.

Houses built from 1976 have walls with insulation standards that were very high compared with those built in the days of cheap fuel, and any benefit from filling the cavity wall with insulation at current price levels will rely on inflation to be cost effective.

Cavity walls built before the '70s may have inner leaves of Fletton bricks or non-insulating concrete blocks, and if so cavity insulation

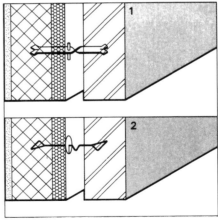

Cavity Insulation. The choice in an existing house is between foam injection, or blowing insulating granules into the cavity. If you are building a new home choose the best of both worlds by building insulating slabs into the cavity, kept in place with special wall ties as shewn. This gives the extra insulation and retains the traditional cavity to avoid any risk of damp penetration.

will offer very real savings. If you are uncertain of the material of the inner skin of your walls, get someone who knows what he is doing to take out a picture hook, and to drill a small hole behind it with a masonry drill. The dust from Fletton bricks, dense concrete, aerated concrete or clinker block can be easily distinguished, the hole plugged and the picture hook replaced. It is most important to have this sort of assessment made by someone who can give impartial advice.

There are a number of competing cavity insulation techniques, and recent legislation has weeded out the unsatisfactory ones. Installers now have to be registered. Nevertheless, a phone call to the Building Inspector's office at your local Town Hall will enable you to confirm whether or not your Council requires you to make a Building Regulation application for this work, and whether the contractor you propose to employ is on the approved list.

Houses built over fifty years ago will probably have solid walls, and their insulation will depend on thickness, and the material from which they are built. The way to improve their insulation is to line them with insulating board, held out from the wall on battens, but to do this successfully may involve a careful appraisal of the damp-course. Local practice in lining solid walls varies in different parts of the country, and you should take the best local advice — starting with personal recommendations from those who have had this work done to their satisfaction.

All dwellings with a high level of insulation and draught-proofing have a built in potential condensation problem. Ventilation to the roof space becomes essential, and the occupants must learn to open windows as appropriate. This may be a chore, but it is the price of progress. The N.H.B.C. publish an excellent booklet on this which all registered builders are supposed to give those who move in to modern insulated homes.

Having satisfied yourself that your house has an appropriate level of insulation to all the circumstances, we now move on to consider the pros and cons of a heat pump in your own home.

CHAPTER FOUR

A HEAT PUMP AND YOU

The requirement for any heating system, or any up-dating of a heating system, is that it should give enough heat, in a way that is convenient, at an acceptable cost. Heat pumps can certainly give enough heat if they are sized for the job. They are convenient if there is somewhere suitable to put the absorber. They will give an acceptable balance between their high capital cost and their undoubted economy in certain specific circumstances. It is these latter circumstances that have to be carefully considered. At the present price levels for alternative fuels, and making a realistic appraisal of how these prices are likely to move in relation to each other, these circumstances are at present —

- **That the building cannot be heated with gas.**
- **That for reasons of convenience or availability, it is not intended to heat it with solid fuel.**
- **That the heating requirement in the building is such that there is a big enough saving to be made by using a heat pump to justify its capital cost.**
- **That the capital cost involved, which in 1981 may be £3000 to replace an existing appliance, or over a £1500 excess over the cost of an alternative in a new building, is within the householder's means, and suits their general financial arrangements.**

In practice this means that at todays costs heat pump are installed in larger homes where a fully automatic system is required, and which are situated where gas is not available. The bulk of such houses and bungalows have already got oil-fired central heating of some sort or another, or have old solid fuel systems which need up-dating. This is where heat pumps are finding a ready market.

The arguments for their use in new buildings are even more convincing, but there is an almost universal reaction to innovation in new housing except where a property is built for a client who insists on a new feature which particularly interests him. The reasons for this take us into the subconscious mind of the house-buyer, who is taking a giant leap into the unknown by buying a new home, and who wants it to be reassuringly orthodox. At any rate, the advice given to developers at house marketing seminars is to attract buyers with talk of something new, and then sell them something familiar. All beyond the scope of this book, except to justify the prediction that if you want a heat pump in a new home, you will have to make special arrangements to get it — at any rate for a year or two yet. It is interesting to recollect how many modern features in new housing were introduced by being sold as replacements in existing houses. Waste disposers, double glazing, even central heating itself was introduced in this way.

If you are in a situation where a heat pump may be appropriate to your needs, the first essential is to arrange for a qualified heating engineer to work out your heating requirements, and to advise on how to meet them. In an existing building this involves working out heat losses through walls, roof, and floor, taking into account many factors including existing and possible future insulation and the size and condition of existing radiators. It may not be possible to put a new heat pump in the same place as an existing boiler, or it may be a good idea to retain the boiler, and allowance will have to be made for the implications of putting the heat pump somewhere else. Eventually you will end up with a heating requirement figure, and often a suggestion that it can be reduced by modest expenditure on further insulation. Then, and only then, are you in a position to look at what the heat pump manufacturers have to offer.

Getting this survey done is not always easy to arrange, particularly if you are looking for advice as to whether or not a heat pump installation is really justified. There are two separate approaches. Either you can ask a heat pump manufacturer to make suitable arrangements, in which case they will either send one of their own technical staff or else arrange for their local installer to handle it. The advice will be free, and you will have to depend on the repute of the people concerned in acting on it. Obviously you will be under no obligation, but remember that those concerned are in business selling heat pumps! This works two ways; not only will they try hard to sell you one if it is appropriate to your needs, but you can assume that a reputable firm is equally anxious not to damage its reputation by selling you equipment that is not suitable for your needs.

The alternative is to retain an independent heating engineer to advise you. He will require to be paid, and consultants' fees can be unexpectedly high, so get a quotation first. The problem is that you want a heating engineer who has wide relevant experience of heat pump installations, and this narrows the field. Neither the Yellow

REPORT ON   for Mr.

Property: A detached 5 bedroom farm house, approx. 100 years old,
                       with thick solid walls.

Heating system: Radiators installed 16 years ago with 100,000 BTU oil boiler.
                       Hot water heated summer and winter by boiler.

Insulation: 150 mm. ceiling insulation and partial double glazing
                       installed 4 years ago.

Oil consumption: Approx. 1400 gals. per annum over last 3 years.
                       Est. 400 gals. for hot water, 1000 gals. heating.

Cost at current price: 16.5 p/litre = £1,051

Heat losses: Not calculated.

Heat consumption: Heating 1000 gals. = 4546 litres at 65% efficiency at
                       37,700 KJ/litre = 30,958 kWh.
                       Over 230 days of heating season ≈ 135 kWh/day ave.
                       Currently heated for 14 hrs. daily = 9.6 kW ave.
                       Peak heat requirement (ave. x 1.6) = 15 kW.

Balance point: (See attached graph) = $3^{\circ}C$.

Percentage of heating load carried by heat pump = 88%

Hot water: 34 kWh/day will heat 116 gals. hot water through $100^{\circ}F$.

Cylinder size: 60 gals. ≈ therefore 50% on day rate, 50% on night rate.

Savings Estimate: Oil required, 12% of 4546 l. = 545 l. @ 16.5 p.   =   £90
                       Heat pump at $\frac{1}{3}$ night rate (1.69 p/kWh)
                                    $\frac{2}{3}$ day rate   (4.4 p/kWh)
                                    C.O.P. 2.5 = ave. 1.40 p/kWh effective   = £381
                       Hot water at $\frac{1}{2}$ day, $\frac{1}{2}$ night
                                    C.O.P. 2.6 = 1.17 p/kWh. effective   = £145
                                                                              £616

                       Saving at current prices: £435 p.a.

                       Cost (see attached) £3,100.
                       Return at current prices ≈ 14% after tax.
                       Pay back, presuming 15% increase in oil and electricity price
                                                                    = 5.1 years.

Your own heat pump — the sort of engineers report that you can expect from a manufacturer.

QUOTATION for Mr.

		£
To	Supply and deliver 1 No. Eastwood 80, 14/11, 3 phase heat pump - split unit.	2,500

To Positioning of base unit in cellar, adjacent to existing oil boiler, evaporator outside back door in agreed position.

To Connecting above with refrigeration lines.

To Connecting heat pump into existing 1" copper heating circuits by manually operated 4 port valve, including supply of all necessary pipe work, valve and additional circulating pump for primary hot water.

To Connecting heat pump to 3 phase, 20 A. TP & N isolator, situated within 1 m. of base unit.

To Testing and commissioning of entire installation.

All for the sum of: 600

£3,100

Guarantee:	12 months parts and labour as attached specimen guarantee.
V.A.T.	Zero rated - extension of existing heating system.
Delivery:	3 weeks from receipt of firm order (Confirmation of Order attached).
Installation:	Requires approx. 2 days - central heating will be off from 9.0 a.m. to 5.0 p.m. on first day.
Payment:	10% with order, balance net on commissioning.

Your own heat pump — the sort of quotation which would accompany the report on the opposite page.

Pages nor the Professional Associations are likely to help here, and the easiest way of finding the right consultant is to ask a heat pump manufacturer to suggest someone. They will know who are the specialists better than anyone else, and will not be in the least disconcerted by such a request.

As with much else, personal recommendation is invaluable in this sort of situation. If you cannot get it, it is at least sensible to ask a manufacturer or installer to take you to see other heating installations for which he has been responsible, and to ask that it should be one of their run-of-the-mill jobs, and not a show installation.

Finally, whatever recommendations are received, insist that any quotation clearly sets out the performance which the installation has to give, the guarantee period, and long-term maintenance facilities. A full years warranty to take a unit through a full season is obviously desirable, and a maintenance man on "same day" call is a good idea however reliable the equipment.

So much for costs. Another factor to consider is what happens in very cold weather. The efficiency of a heat pump falls off as the temperature drops, at the time that the demand for heat in the house is at its greatest. A heating engineers report will detail the balance point when the output of a specific heat pump drops below the heating demand at that particular outside temperature. This is usually around freezing point, and for the technically minded this is discussed further in Appendix 1. The average daily winter temperature in Britain is 43° F. It is only on about ten days a year that the average daily temperature falls to 30° F, which means that it is freezing throughout the twenty-four hours. If the balance point of the heat pump proposed is at 30° F, this means that there will be ten days in the average year when you will either have to turn off the radiators in the spare rooms, or use some other source of heat, such as an open fire or an electric appliance. If you have retained an existing boiler, this is the time to use it. And if you don't like the sound of this you can always opt for a heat pump that will give a lower balance point in the system, or else a model with a booster system, that provides extra heat when required.

Booster heaters that depend on full tariff electricity are very expensive to run, and can only be used for relatively short periods if their electricity consumption is not to be noticeable. If they are designed to be turned on automatically, and anything is amiss with

Heat pumps are new. If you cannot choose between those available on personal recommendation it may be a good idea to visit the factory where they are made. The photographs show heat pumps awaiting despatch and a unit on test at Eastwoods factory near Mansfield.

the controls, this could be very expensive indeed. A situation where this happened is described in Case History 6. In all normal circumstances it is preferable that booster heaters should be switched on manually, and that there should be some form of warning light to show that they are in use.

The running costs of a heat pump are not only a matter of how much electricity it uses, but also depends on the cost of the electricity. Different Electricity Boards have different scales of charges, and these are set out in Appendix 2. The differences between them are not significant, but the differences between the various tariffs are enormous. There are three basic scales of charges for selling electricity to private consumers: Full Tariff, White Meter Tariff, and Economy 7 Tariff.

Full tariff electricity costs the same for all twenty-four hours of each day, and this is the basic rate which applies to most households.

The White Meter tariff involves paying different prices for electricity used at different times of the day, and is no longer available unless you already have White Meters installed.

The Economy 7 tariff* is the one which is best suited to a household with a heat pump. It involves paying up to 10% more than standard tariff for units used during seventeen hours of the twenty-four, and 40% of the standard tariff for units used during the other seven hours, which is usually between midnight and 7 a.m. A few minutes with a pocket calculator will demonstrate that as long as you use more than 10% of your electricity at the cheap rate, the Economy 7 tariff gives cost savings. A heat pump installation typically takes one third of its units at the cheap rate, and this is clearly demonstrated in the Case Histories. If a heat store is used, the savings are higher still.

These savings depend on the controls being programmed to use cheap rate electricity to heat the day's hot water and to bring the whole house up to temperature in the early morning. This is discussed in a later chapter.

The Electricity Boards do not usually charge for replacing standard tariff meters with Economy 7 meters but you may be involved with an Electricity Board charge for a 3-phase electricity supply. A normal domestic electricity supply is "single phase", and if you have only one electricity meter, you have got a single phase supply. Most heat pumps will work on single phase, but some of the larger

Still called White Meter Tariff in Scotland

Nearly all Heat Pump installations take advantage of Economy 7 tariffs. Different Electricity Boards have different scales of charges, which are summarised in Appendix 2.

ones prefer a 3-phase supply. This involves a supply cable with extra wires in it, and extra meters. If you have three meters already, and your electricity bill is for the total units recorded by the three meters added together, you already have a 3-phase supply. If not, you may be advised to consider making a change. The cost of this can vary from nothing to several hundreds of pounds, and this is entirely at the discretion of the Electricity Board. Do not worry too much about this: the heat pump manufacturers are very conscious of the 3-phase electricity problem, and it is probable that all new models will be available with "soft start" equipment that enables them to work on single phase. However, three-phase electricity gives marginal cost savings, and if you already have three-phase wiring, or can have the change made at a nominal charge, it is very worthwhile.

The next consideration in discussing the installation of a heat pump is deciding where to put it, and this leads to a decision between split and combined models if you are considering a unit from a manufacturer that offers the option. Both are shown in the photographs, and explained in the captions.

Combined models have all of the machine in one housing. If they are units that collect heat from the air, this means that they will normally have to stand out of doors in a fairly exposed position.

Split models have a heat absorber standing outside, and the rest of the machinery in an attractive cabinet intended to be put inside the house. There is little advantage for this: the reason for it described to me by one manufacturer is that the customer expects to have his heating appliance in the house, so this is what is offered. Another reason is that heat pump fashions have imitated the Germans, who are well ahead of Britain in installing domestic heat pumps. In Central Europe heat pumps are put indoors because they are used to take heat from the ground, and there is no point in putting them outside. If they take heat from the air, then the air is ducted to them to avoid snow drifting over the machine. In Britain there is no reason why one should not opt for a combined unit, with all the machinery in a weather-proof casing out of doors. This will save space in the kitchen, utility room or garage, which is where space is usually at a premium.

Siting a combined unit or the heat absorber of a split unit requires that it should have a good air circulation, should be clear of places where snowdrifts collect, for the pipe runs to be reasonably short,

Above are the two parts of a split model, below the same manufacturers combined version of the same heat pump. In most situations the latter seems a more sensible arrangement.

and for it to be put where the noise it makes will not be a nuisance. Heat pumps are certainly no noisier than pressure jet oil boilers, but they can be heard working, and this must be kept in mind. Another factor to remember is that the heat absorber part of the apparatus drips water which has condensed on the cold absorber coils. This is quite normal. In the winter this water freezes, and all heat pumps incorporate automatic de-frosting arrangements which melt off this ice from time to time. This condensate, which can amount to several pints an hour, is either allowed to find its way into a grass surround, or else is collected and run into a small drain. Its steady dripping will certainly lead to visitors reporting that your heat pump is leaking.

A new heat pump installation also requires consideration of the sizes of existing radiators. Heat pumps require rather larger radiators than oil or solid fuel systems. Radiators installed in a new house will be of a size to suit the system, but in an existing house it is obviously necessary to use the existing radiators if this is possible. Fortunately it is, as virtually all houses have radiators that are over-sized due to the standard of insulation in the building having been improved over the years. It is unusual for them to be changed to suit a new heat pump, but this is something that will be checked by the heating engineer who advises on the whole installation. If a change does have to be made, it is generally possible to fit a new high output radiator of the same size as the old one, so that no alterations are required to the pipe work. Incidentally, very old cast iron radiators, as used with old gravity heating systems in period houses, work very well indeed with heat pumps, provided that the piping can be properly insulated.

While existing radiators rarely have to be changed, consideration should always be given to changing the domestic hot water cylinder, unless the existing one holds at least 40 gallons. The advantage of heating as much hot water as possible at the cheap night rate is obvious, and the only limit to the size of the hot water cylinder should be the weight that can be taken by the floor joists: a fifty gallon tank weighs a quarter of a ton, and is the usual size. The tank should be very well insulated, preferably by using one of the new type with sprayed on foam insulation.

Finally, in considering all the aspects of installing a heat pump, mention should be made of the use of other heat sources. All of the case histories in this book concern machines which take heat from the air. This is because for 99.9% of potential users in this country the air is the most practicable source. Very few of us have the opportunity to use water for this purpose, but this is discussed in Chapter 7. No other heat source is worth considering in an ordinary domestic situation, but as many alternatives are featured in magazine articles and so on, they deserve mention.

Taking heat from the ground by digging a hole of twice the area of the house and burying a pipe in it works well, and is widely used in Germany, Scandanavia, and other places that have very low winter air temperatures. This has been tried in Britain. It is not significantly more efficient than an air source installation, and if you feel like digging a hole twice the size of your house, this may be of interest to you. If not, read on. Putting a heat absorber in the roof to make use of waste heat from the house and to benefit from the warming effect of the sun on the tiles has been widely advocated. It does realise both these advantages, and an installation of this sort is described in Case History 7. This arrangement certainly does away with the problem of where to put an absorber in ones garden, but it involves special ducting in the roof which may be both difficult and expensive. One also has the problem of getting the thing up there. The practical experience of manufacturers and installers has led them to steer clear of this particular idea, however attractive in theory.

The same is true of arrangements linking heat pumps with solar panels. There are many special considerations here, and the practical problems involved make theoretical advantages irrelevant. If shortage of space or other factors makes a standard system impossible, then a solar/heat pump arrangement may be worth looking at, but this would be a venture into the world of experimental installations. Here one hopes the fun will balance the expense, and success is a bonus. Heat pump economics hinge on cheap rate electricity, which is only available during the hours of darkness, when solar panels are not at their best.

Arrangements to extract heat from domestic waste water are often discussed. No simple way of doing this would be legal under existing building by-laws and public health legislation. The practical difficulties would be considerable, the cost non-economic, and the possibility of freezing up the drains would be of interest to your neighbours.

None of this is criticism of those who put forward these proposals, whether in the media or in experimental housing. They do an important job, but in this book we are concerned with proven heat pump technology that is commercially available now. It is essential to differentiate between the two approaches.

Finally, remember that as a tax-payer you pay for investigations into Heat Pump efficiencies by several statutory bodies. The best place to get reports on this research which you have financed is the bookshop at the Building Centre, Store St., London. As your tax pays for the Building Centre as well there is no need to be shy about using its services.

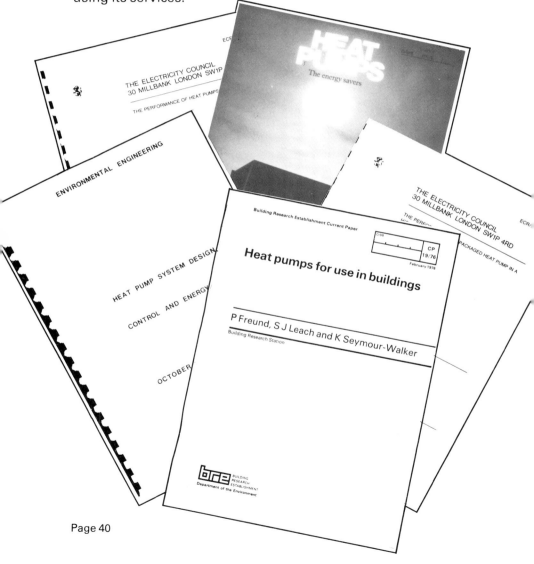

THE ELECTRICITY COUNCIL
30 MILLBANK LONDON SW1P

THE PERFORMANCE OF HEAT PUMPS

HEAT PUMPS
The energy savers

ENVIRONMENTAL ENGINEERING

HEAT PUMP SYSTEM DESIGN

CONTROL AND ENERGY

OCTOBER

Building Research Establishment Current Paper

CP
19/76
February 1976

Heat pumps for use in buildings

P Freund, S J Leach and K Seymour-Walker
Building Research Station

THE ELECTRICITY COUNCIL
30 MILLBANK LONDON SW1P 4RD

THE PERFO

PACKAGED HEAT PUMP IN A

ECR/

bre BUILDING
RESEARCH
ESTABLISHMENT
Department of the Environment

CHAPTER FIVE

CONTROL SYSTEMS

Controls for domestic appliances have become progressively more sophisticated in recent years, and technical developments, including the inevitable micro-chip, are making the cost of control equipment a relatively small part of the cost of the appliance itself. The heat pump is a very expensive machine which will heat a house very cheaply. Further modest expenditure on controls to enable it to do its job as efficiently as possible is essential. In this chapter the various possibilities are examined by looking at the different pieces of equipment that are available, and to keep things simple they are considered as they might be applied to an air-source heat pump supplying hot water to an ordinary family house with radiators and a hot water cylinder.

The first control gadget is free to those who opt for an Economy 7 electricity tariff. When the meter is installed the Electricity Board's engineer can be asked to provide a "fourth tail", or wire that will only be live when the cheap rate units are being counted by the meter. Your own heat pump installer can use this signal to open an automatic valve to heat the hot water cylinder, or to start putting heat in a heat store. At the end of the cheap rate period the same arrangement will disconnect the cylinder or heat store.

An ordinary time clock under your own control can determine when anything at all happens: the time when heat is available to radiators, or the cylinder, or when the heat pump should be closed down altogether, or that it should start up for half an hour every day to keep the house aired while you are on holiday. The essential thing is that it should be able to switch on and off as many times as your system requires, and, unless you never go away for weekends, that it should be a seven-day model. The difference in cost between 7-day time clocks and cheaper twenty-four hour ones is very little, and the former have many practical advantages.

Thermostats are temperature sensors which are used to turn equipment on or off at pre-set temperatures. In a heating installation they can be used to turn the heat pump itself on or off, or to control the water circulating pumps, or to open or close automatic valves. One of the most important in the system is the familiar wall thermostat which controls whether or not the circulating pump is pushing hot water around the radiators. This is traditionally put in the hall, usually near a stairwell, which is the worst possible place for it. Halls, and stairwells in particular, get a lot of air changes as people move around the house and a wall thermostat in this position is unlikely to be sensing the temperature in the living rooms. If you have any choice in the matter discuss the position of your wall thermostat with your heating engineer.

It is becoming common to split the piping circuits in a house so that the radiators upstairs are controlled separately from those downstairs. An arrangement of this sort involves the first floor having its own thermostat, possibly linked with a separate time clock.

Digital thermostats arrived in 1980, and one is shown on page 46. This has an automatic "set back" feature, with a timing device that automatically sets back the temperature setting at night by a few degrees. This is a useful feature for heat pump users who want to keep their heat pump running at night to keep the house warm, but do not want it at a full daytime temperature.

Another thermostat is the cylinder thermostat which senses whether or not the hot water storage cylinder requires heat. Most of the older oil-fired installations are arranged so that the cylinder is heated whenever the boiler is working, whether or not this is necessary. If your cylinder does not have a thermostat, ask your heating engineer whether it is practicable to fit one.

Wall and cylinder thermostats are designed to be adjusted by those living in the house to suit their own needs. There are likely to be others in a heat pump system, both in the pump casing itself and outside it. Some may be clamped to pipes to sense return water temperatures in order to control automatic valves: others will be safety devices. They will be adjusted by the heat pump installer, and if children are liable to experiment with them the dials should be taped up.

Thermostatic radiator valves turn individual radiators on and off automatically, and can be adjusted like wall thermostats. They are relatively inexpensive, and provide a very cheap and effective way of ensuring that you only heat rooms that you want to heat, and only let them reach the temperature that you require. Typical uses are to take the chill off a spare bedroom that is not in use without heating it, or to control radiators in a room where there is an open fire. The

Control System Components
Top — seven day time clock
Left Centre — cylinder thermostat
Right Centre & Below — thermostatic radiator valves

two different types of automatic valves shown in the photographs are in the drawing room of the author's home, and turn off the radiators if the fire is lit in the evening. The two different types were installed so that their performance can be compared: there is nothing to choose between them.

Frost-stats are simply thermostats designed to over-ride all other controls and start up a heating system when there is any risk of the water in any part of it freezing. Some heat pumps incorporate frost-stats inside the casing: it is important to find out the automatic frost protection built into your system, if any.

Control boxes receive the signals from time clocks and thermostats, analyse them, and send the necessary switching instructions to the heat pump, circulating pump and automatic valves. Some are obviously designed for eye-appeal and look like the controls for a music centre. One of these will impress your friends, and this may be worth the extra money. All of them work well, and are reliable. Unless space is at a premium do not buy a very small control box unless your electrician has very small hands to make the connections. In general it is best to use the equipment recommended by the heat pump manufacturer.

All heat pumps have some of this control gadgetry "built in", and some come with most of the rest as a package. Designing your own control system is not recommended for anyone but a systems engineer as it is very easy for the amateur to get a control system oscillating and continuously switching itself on and off. On the other hand, as you are going to live in the house, and as you are paying for the installation anyway, you want to control how you live and not leave it to anyone else. The thing to do is to draw up a list of what you want your heating system to do, and then tell your heating engineer to arrange your control system accordingly. The check list opposite may help.

CHECK LIST FOR THE CHOICE OF AN AUTOMATIC CONTROL SYSTEM FOR A HEAT PUMP INSTALLATION WITHOUT A HEAT STORE

1. Are you using the Economy 7 tariff? If so,

   a) is the hot water cylinder to be heated at cheap rate?
   b) is the house to be kept warm all night at cheap rate?
   c) is the house to be brought up to full temperature at cheap rate before the rates change at 7.30 a.m.?

2. Do you want two different heating zones in the house, usually one downstairs and one upstairs?

3. Are thermostatic radiator valves relevant to your requirements?

4. Do you have a cylinder thermostat, or can one be installed easily? (If not, your cylinder heating can be controlled by a sensor on your return flow pipe.)

5. Where do you want the time clock situated? Do you go away for odd days or weekends and need a seven-day time clock?

6. What frost protection does your system require?

7. If your heat pump has a "back up" auxilliary heater of some sort, do you want to turn it on manually or automatically? Do you want the best of both worlds with automatic control once you have manually selected this facility?

8. Do you want a warning light in the kitchen to tell you that the auxilliary heating is being used?

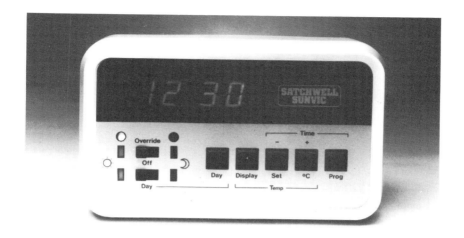

Above — digital wall thermostat with "set back" feature particularly convenient for controlling heat pump installations.

Below — Control Box developed for controlling heat pumps by Church Hill Systems.

CHAPTER SIX

HEAT STORES

The availability of off-peak electricity tariffs, with electricity consumed between midnight and dawn costing less than half price, makes some sort of heat store an attractive proposition. If heat from a heat pump working at night at a special price can be stored until it is required during the day, it offers the possibility of tremendous savings, and, as already discussed, will give the lowest running costs of any system, whether it is oil, gas, solid fuel or electric. This is using the principle of the familiar electric night storage heaters, and of the Enviwarm and other electric warm-air central heating systems that are essentially large storage heaters serving a whole building. The trick is to store the heat effectively, with minimum losses, and to retrieve it from the store efficiently.

Heat stores to work with a heat pump are already a practical proposition, but here again consideration of cost-efficiency, convenience and aesthetics dictate whether or not they are a commercial proposition. The electric night storage heaters which have been with us for thirty years, use iron ingots or clay fire bricks to hold the heat generated by electric heating elements embedded in the stack of ingots or bricks, and in this way are heated to a high temperaure. When the heat is required, it is extracted by blowing air through the hot stack. This air is filtered and then fed out to wherever it is required. It is difficult to imagine a heat pump linked effectively to a store of this sort, but it is both practical and economical to use water to store heat from a heat pump, and most of the installations made today incorporating heat stores are wet systems. However, this situation may change very quickly, as new heat storage techniques based on phase change materials are being very actively developed.

Phase change materials used in heat stores are chemicals which either give up heat, or absorb heat, when they change state from solids to liquids and vice versa. This is called the latent heat of fusion, and can be harnessed by putting heat into a container of solid phase change material, chosen to melt at a useful temperature. It will absorb a great deal of heat in the process of melting, and when the the process is reversed, and it is allowed to solidify again, it will give up heat at the same useful temperature.

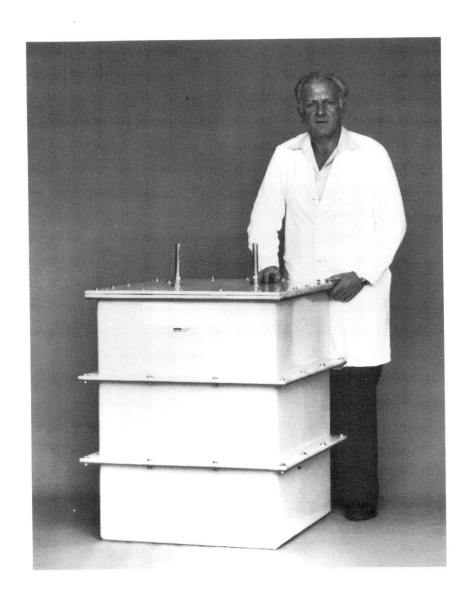

Calor 31 heat store using Glaubers salt with a 60 kWh capacity — virtually the same capacity as the authors large water filled heat store shewn in the photograph on page 54.

There are a great many materials that have this happy knack, and the advantages of making use of them are overwhelming. One of the more common ones is Glaubers Salt, Sodium Sulphate Decahydrate, and a heat store using the latent heat of fusion of this material need have only one third of the weight and one fifth of the volume of a comparable tank of water which would hold the same amount of heat without a phase change. The photographs of water heat stores in use today compared with those of a prototype phase change heat store demonstrate clearly the size advantage. However, at present this is still a very new concept, and presents engineering problems of its own. Specially formulated chemicals with specific phase change temperatures are expensive (They are usually hydrated salts of alkaline metals, or else organic waxes). They are liable to change their volume at the phase change, often have a poor thermal conductivity, and may be corrosive. This is incidental: the phase change heat store offers such advantages that the necessary investment in research is being made, and compact and convenient heat stores must become commonplace in due course. However, this is in the future.

In most circumstances today a heat store linked to a heat pump will be a water tank, invariably directly piped so that the water in it circulates either to the heat pump or around the house, as controlled by automatic valves. During the off-peak tariff the heat pump heats it up to the maximum possible temperature, and during the day the heat demand from the house is met by first using hot water from the store. This may either circulate through conventional radiators, or may be used in a warm-air system. Either way, automatic controls will ensure that the heat in the heat store is used first, and that the heat pump cuts in automatically when it is exhausted.

In arranging all of this the size of the heat store is critical: too small and it will not enable the best use to be made of the off-peak tariff; too large and it will not get up to the optimum temperature in the off-peak heating period. For someone considering a heat pump today, the heat store calculations to be made are generally as follows:—

Off-peak tariff	7½ hours per night
Proportion of this required to heat domestic hot water and to give an hours heating directly into the radiators in the early morning say,	1½ hours a night
Leaving output from the heat pump available for the heat store	6 hours a night
Heat store to be heated to	140° F.
Temperature of heat store after usable heat has been extracted from it during the day.	110° F.
Therefore heat store has to be heated through	30° F.
Output of heat pump at freezing point for six hours	48 kilowatts
Quantity of water that 65 kilowatts can heat through 40° F.	2233 litres (491 gall)
90% of this providing a margin to ensure that the store always gets up to temperature	2000 litres (440 gall)

The issue is then to consider the saving in having nearly 50 kWh of heat available for use during the day against the capital cost of providing the heat store tank, installing the pipe work, valves and control gear, insulating the whole lot, and, often the most difficult thing, finding somewhere to put this inconveniently large tank and some way of disguising its unattractive appearance. A complication in this is that the tank has to stand the pressure of the central heating system, which is the head of water measured from the bottom of the heat store tank to the top of the expansion tank in the loft. Unfortunately, this pressure is too great for a slab-sided tank in light plate, and it is not possible to use a standard 600 gallon oil tank which may be available from an oil-fired system supplanted by a heat pump. A typical 500 gallon tank made to stand 30 p.s.i. pressure will cost in excess of £300, and this is a field where there is endless scope for improvisation. Some of the photographs show heat stores which are simple water tanks; one shows the author's 500 gallon stainless steel glass lined brewery tank bought second-hand for £195. Another in Case History 5 uses a battery of large domestic hot water cylinders with Herculag insulation.

All this has assumed that the water in the heat store will be circulated directly through the radiators, or directly to the heat exchanger in a warm-air system. It is on this basis that consideration has to be given to the pressure in the system. An alternative is to have a heat exchange coil in the tank, which then need not be pressurised. There would be some lack of efficiency in this, and even if the tank was already available the cost of installing the coil would probably be in excess of a new tank to take the pressures of a direct system. However, one complication of a direct system as described is that provision has to be made for the expansion of such a large volume of water. An allowance of 5% should be made. Many central heating systems with oil or solid fuel boilers have header tanks of only 10 gallon (45 litres), and the cost of replacing these is a further factor in the heat store cost equation.

Old Brewery tanks, which are often made from stainless steel and glass lined, make excellent heat stores and are available cheaply. These are at the yard of Central Bottling & Brewing Services Ltd at Bircotes, near Doncaster.

The actual piping and controls for a simple direct heat store are best studied by reference to the photograph of an actual installation on the facing page. The various piping circuits are controlled by five solenoid valves, and the operating sequence from the Church Hill Systems control box is as follows:—

MIDNIGHT
When the Electricity Boards meter switches over to the cheap tariff it automatically opens the valves in the heat store pipe-work, and if the heat pump is not already working it turns it on. The heat pump will continue to put heat into the heat store unless it is also sending heat to both the house and the domestic hot water cylinder, and the controls sense that it is unable to meet all three demands. In this case — which is unusual — the heat store is automatically disconnected.

6.30 a.m.
The valves to the house open and part of the heat pumps output for the last hour of cheap tariff electricity is put into the house. As the heat store is then nearly up to full temperature of 140° F. the pump can cope with this.

7.30 a.m.
When the Electricity Board meter switches back to full tariff it automatically causes valves to shut so that the heat demand from the house is met from the heat store. The heat store may last until mid-day, or until the evening, depending on the weather. When the temperature of the heat store falls to 110° F. it is automatically disconnected until midnight, when the cycle starts again. However, if while the heat store is being used the hot water cylinder calls for heat, the heat pump is switched in to meet this requirement as the cylinder wants hotter water than the heat pump may be able to supply.

The radiators in the house are controlled in three ways; by their own thermostatic radiator valves, by a wall thermostat in the hall, and by a time switch. The wall thermostat and the time switch both have to call for heat together for the system to send heat to the house. The hot water cylinder has a cylinder thermostat, but this is not controlled by the time switch so the cylinder will always be heated whatever the time is.

Controls for the heat pump with a heat store illustrated overleaf. The control sequence is detailed opposite. The oil boiler is the original appliance used before the Heat Pump was installed.

The authors own installation, showing the heat pump to the rear and the heat store in front of it. The lower photograph shews the store, which is a brewery tank, before it was insulated and its wooden housing built.

All this sounds very complicated, but it is all completely automatic. In terms of technology, the control engineering involved is on the same level as that in a modern oven. Housewives who can cope with an oven would soon adjust to the sort of heating control system described. The problems come with husbands who attempt to modify the controls!

There is one final type of heat store which is totally different in concept. This is to heat the whole of the floor of a new house with off-peak electricity, and to let the heat diffuse out into the building all day. This is done by embedding a serpentine coil of piping in the concrete at the time that the building is constructed. Sometimes a wall is treated in the same way. This particular approach to domestic heating is one that received a great deal of attention in the text books, but which is rarely seen except in experimental buildings. Obvious problems are a lack of flexibility, and the prejudice against underfloor heating which undoubtedly exists. However, in new houses where the floor or a masonry wall is to be used as a thermal store, the heat pump becomes an obvious choice as an energy source, provided that the size of the installation justifies the capital cost.

Leaving aside under-floor heating, a major practical aspect of providing a heat store is where to put the thing. The ideal is a cellar, where any heat losses will find their way up into the house, where pipe runs are likely to be short, and where the installation is both out of sight and yet can be inspected easily. An out-building is a good second best. If a heat store has to be in the open, it presents problems with insulation and frost hazards, as well as requiring consideration of how it will look. Heat losses, can be dealt with by installing all the insulation that is recommended — and then some more just to make sure. Frost precautions should include isolating valves and a drain-down facility, with perhaps a frost-stat and appropriate control circuitry to start the heat pump if one is an enthusiastic gadgeteer. The aesthetics must depend on circumstances, and various approaches are shown in the photographs. Wherever the tank is situated, it is essential to provide adequate air-venting in the system to avoid mysterious noises at night. An automatic snifter valve is preferable to an air bleed.

CHAPTER SEVEN

WATER AS A HEAT SOURCE

As discussed in previous chapters, there are theoretically many potential sources of low grade heat which are put forward for utilisation using heat pumps. This book is concerned only with practicable applications of heat pump technology to domestic heating today, using commercial equipment that is readily available, and in ways that are cost effective. In practice this means using either air or water as a heat source. Other sources are either practicable but not cost effective, or simply impracticable for general use. They are discussed briefly in chapter four.

Heat pumps that use water as a heat source are simpler and thus cheaper than those previously considered, as water is both a better conductor than air, and is usually available at a constant temperature that enables higher efficiencies to be achieved. Unfortunately there are invariably practical aspects that tend to dominate consideration of this type of installation. One has to arrange for three things — enough water, the right water, and for it to be in the right place.

If the water is to be pumped or chanelled to the heat pump the requirement is quite simple: a flow of 50 gallons per hour at at least 8°C for each kW of output required if the supply is from a constant temperature source such as a spring, a well, or waste industrial water. At least twice this flow is required for water from a river, lake or the sea, which may be at a lower temperature in mid winter. Water that can be relied on to be at a temperature of 8°C or above all the year round can be piped into the heat pump itself, and will leave it a few degrees colder, but in no danger of freezing. Water from other sources that may be at lower temperatures can freeze in the heat pump itself, and have to be used to give a flow through a heat exchange sump.

Alternatively, a heat exchanger coil can be situated in a river or large stream, mill race or other artificial waterway. The sea would be even better if it is available.

Unfortunately only a fraction of the large well insulated houses for which heat pumps are suitable are anywhere near a water source of this sort. Very few new houses are in such a situation, and old houses on the banks of waterways are invariably so badly insulated that the cost of appropriate new insulation to the required standard is likely to be unacceptably high unless it is installed as part of a general renovation. However, if a water source is a practicable possibility the first consideration is to establish the flow available. This will probably be seasonal, but here nature is on our side and maximum flow in winter coincides with maximum heat demand.

The availability of the thousand gallons per hour required for a system where the water is taken to the heat pump is easily determined by hiring a pump of a suitable size and simply running it for a week under the worst conditions likely to be met. If your spring, well or stream runs dry, or the pumping reduces water levels unacceptably, you will soon find out. Suitable pumps are available from contractors hire depots in every large town, and the cost is negligible in the context of the total cost of the installation proposed. The actual volume of water delivered by the pump can be measured by timing how long a 40 gallon drum takes to fill and then converting this to a flow per hour.

A far larger volume of water has to flow past a heat exchanger in a stream, and the basic requirement is that the exchanger coil or grid be completely covered to a depth of at least 6" at all seasons, that there should be a clearly discernible flow through and around the coil, and that it should never ice up completely in winter. The installation shown in the photograph on page 59 illustrates full advantage being taken of a small stream, and a smaller watercourse than this could not be used in this way unless the flow is very carefully regulated by weirs and an artificial channel provided in the area of the heat exchanger.

Any arrangement to take water from a watercourse or well, or to put a heat exchanger in a stream or river, will require a license from the Water Authority and probably the consent of others who have rights in the matter. This applies even though you may plan to return the whole of the flow to the river, and the fact that it will be returned at a different temperature will be considered with great care.

Having the right water is a different matter. Most wells and many streams in Britain are at a constant temperature of about 10°C. The sub-surface flow in deep rivers is usually a little colder, while surface water and shallow streams freeze at the very time that the heat pump is required to work hardest! Some water is corrosive, other water leaves scale in pipework. If corrosion and scaling are avoided by using heat exchangers in the water instead of taking the water through pipes, the water source may respond by encouraging algae growth on the exchanger coils, or simply clog them with water borne debris.

This leads to the fact that a heat pump installation has to be managed, and requires a certain amount of interest taken in it. Consideration of a water source system pre-supposes that the water intake and discharge points will be inspected at intervals, and that heat exchanger coils in the stream are kept clear of debris. This is far less maintenance than is required by, for instance, a swimming pool but far more than is required by a gas fired central heating boiler.

All of this makes water as a heat source a much more practical proposition for those who have already got an existing water supply under their own control. Fish farmers, farmers and horticulturalists who use water for irrigation, or who have elaborate land drainage arrangements, are in this category. It is likely that they have existing spare pumping capacity to move 1000 gallons per hour to the heat pump, and it may even suit fish farmers to have cool water to return to their ponds.

A mill stream controlled by a weir can be an ideal heat source for a heat pump installation in a converted mill, and the possibility of combining two technologies and using the mill wheel to generate electricity to drive a heat pump is an irresistible concept for the technically minded purchaser of an old water mill. In environmental engineering terms this is a very good thing to do: in economic terms the capital costs involved make the idea strictly something for those who can afford to let energy conservation be their hobby, or for those with the ability to make their own hydro generating equipment at very low cost. To prove the point, a mill race with a 8'0" head requires a flow of 4500 gallons per hour to generate 5 kW of hydro electricity. Commercially available equipment to harness this currently costs many thousands of pounds before it is installed. Add to this two thousand pounds for the heat pump and another one thousand pounds installation cost and it becomes cheaper to spend the winter in the West Indies and avoid heating bills altogether.

Returning to more practical propositions, a pumped water source installation does offer an opportunity to pick up waste heat from other places before water gets to the heat pump. If the water feed

Two water source heat pumps which provide full heating and all the domestic hot water for a large new house in the E. Midlands. The top photograph shows the absorber coils in the stream at the bottom of the garden.

pipe is run through a coil in a slurry tank on a dairy farm, or under a stockyard floor, it can do no harm and may add significantly to the efficiency of the system. All will depend on the cost of any such arrangement, but it is likely that work of this sort will be done by the property owner concerned and the real cost to him may be negligible. If waste heat from commercial or industrial premises can be used in this way the savings could be very significant. Certainly no arrangement for warming water fed into a heat pump can possibly do any harm and provides an opportunity for the conservation enthusiast to let his imagination and igenuity run riot, while on the other hand any unorthodox arrangement for putting the absorber coils of a heat pump directly into an unusual heat source requires very careful professional assessment.

A professional assessment is the key to the whole business. A reliable water source provides constant conditions which will enable a design engineer to specify equipment which will be significantly more efficient than air source equipment. Unfortunately the very small number of situations in which installations of this sort can be made in this country makes finding someone with directly relevant experience something of a problem. If you are seeking the advice of a manufacturer's representative it is probably worthwhile enquiring firmly how many installations of this sort he has dealt with before, and asking to be taken to see them.

CHAPTER EIGHT

HEAT PUMPS & SWIMMING POOLS

The lavish use of fossil fuels to heat private swimming pools is becoming prohibitively expensive to all but a favoured few, and is already illegal in parts of the U.S.A. where new legislation requires that only low energy heating appliances can be used for this purpose. The capital cost of a typical outdoor private pool can be matched in only three years by the cost of using oil to heat it to the temperature appropriate to the luxury image which a pool projects. If a pool is not heated to this temperature, it is unlikely to be used as fully as the owner would wish, and becomes an irritating reminder of unrealised ambition.

One answer to this lies with the heat pump, which is even better suited to heating swimming pools than it is to heating houses. Consider the heating requirements: this is to warm a large volume of water, at night, for the part of the year when night time ambient temperatures are at a maximum. Given that a pool cover will be used to reduce heat losses, the heating requirement of most domestic outdoor swimming pools can be met by either a small and simple heat pump which warms the pool water as it passes through the filter system, or else by using a domestic heat pump in an adjacent house which will have spare capacity in the summer, when it does not have to provide central heating. The choice between the two options will be determined by practical considerations such as the distance of the house from the pool, the position of the filter room, and the ease with which new piping can be installed. Small pool heat pumps cost about half the price of the smaller domestic units, and are less sophisticated. They are designed for use in the swimming pool season only, and most lack the costly de-icing arrangements of domestic models, simply switching themselves off and waiting for ice which has formed in the heat absorber to melt to water. Because they work at relatively high ambient temperatures they are more efficient than the heat pumps used in a central heating role, and usually feed about 4kW of heat into the pool for every kW of electricity used to drive the machine. It is usual to connect them into the filter piping and to use the filter pump to push the pool water through the heat pump itself. They are maintenance free, and have waterproof cabinets which enable them to stand outside, saving the cost of a boiler room. Dispensing with a boiler room can be useful when designing a pool that is to be

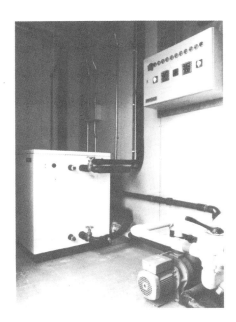

Typical heat pump installation in a swimming pool filter room.

Dehumidifying equipment linked with the heat pump at the same pool.

a feature in a garden.

If a domestic heat pump is to be used in a dual role, providing central heating in the winter, and warming a pool in the summer, it will have to be used with a special heat exchanger. For various reasons it is not a good idea to pump pool water through a domestic heat pump. Although the circulating arrangements for the pool water filter can be used to move it to the heat exchanger, the water circulating between the heat exchanger and the heat pump will need its own circulating pump, and the necessary controls will be more complicated than for a simple pool unit. However, the efficiencies obtained at different ambient temperatures will be much the same, and the advantages of using the same machine right through the year are overwhelming. In practise, while these advantages are recognised, pools are usually sited to suit aesthetic or other considerations, and it is often inconvenient to connect them to the heating system in the house. As a result the simple pool heat pump is more usual.

Heating indoor pools is a different matter, and requires special consideration. Enclosed pools use energy in three ways: to heat the pool water, to heat the air around the pool, and to cope with humidity by either providing artificial ventilation or working a de-humidifier. A heat pump will heat an indoor pool in the same way that it heats one outdoors, and with at least the same efficiency. It can also heat the air around the pool if this is required, and, most usefully, can be used to de-humidify the air at ceiling level, returning heat to the pool and reducing the need for ventilation. Designing an installation of this sort is a highly specialised matter, and expert advice should be sought from the companies or consultants with directly relevant experience.

CASE HISTORIES

These case histories are based on visits paid in February and March 1981. They are the real stories of the involvement of real people with heat pumps, and all that follows has been checked by the individuals concerned. The author is grateful to them for their generous help.

THE 1957 FRIDGE-HEATER

In this first case history we look at a machine that was bought by a Gloucestershire family at the Building Exhibition at Olympia in 1958, that has worked hard and well for 23 years, and which demonstrates clearly the theory, the potential and the utility of the heat pump. Like many brilliant innovative ideas, it was not a commercial success.

The Ferranti Fridge-heater was marketed from 1955 to 1959 and was a simple heat pump for installation in a larder, with pipe connections to the hot water cylinder. It took heat from the larder, cooling it to a level where it was effectively a refrigerator, and transferred the heat to the hot water cylinder, where it would meet normal family hot water requirements. It would even make ice. The cost in 1958 was £130, and the electricity consumption was a maximum of 10 units a day. Several hundred Fridge-heaters were sold, but the appliance never really caught on. This was probably due to its own efficiency: it needed a minimum of 120 cubic feet of larder to keep cool in order to avoid what the instruction book called "the risk of experiencing very low temperatures". This was at a time that modern fitted kitchens were beginning to replace old-fashioned larders, and as the fridge-heater was essentially something for new houses, it did not really have a hope of achieving the sales necessary to sustain volume production. No-one was likely to design a house with a huge brick-built larder simply to be able to install a Fridge-heater, and those who were building new homes that had got big larders anyway were likely to belong to the more conservative and least innovative section of the community.

There are a number of these machines still working — they get talked about by refrigeration engineers, and in the trade there is usually someone who knows someone who knows where a Fridge-heater is still working. The one that the author found was mentioned to him by an architect, and was happily working away when inspected. It is in a larder of the minimum recommended size — a far larger larder than is usual in modern homes — and was installed when the house was built in 1958. The airing cupboard and hot water cylinder are directly above it on the first floor, keeping pipe-runs to a minimum. It meets all normal family hot water requirements, although there is an immersion heater which is

brought into use if more than three baths are required in quick succession. In 23 years of continuous operation it has proved very reliable, although at one stage the fan failed, and was replaced by a local engineer. It keeps the larder as cold as the bottom of a modern fridge, but does not get nearly cold enough to store frozen foods. It will make ice cubes, but is not used for this purpose. Its only show of temperament comes in hot weather when, if no-one wants a bath and the hot water cylinder is up to temperature, some of the hot water has to be run to waste to keep the fridge side of the installation working.

The Fridge-heater was not only the first appliance of its type on the market, it is still the only commercial heat pump ever sold in Britain for domestic use which used the temperature differentials at both sides of the system. With micro-chip technology bringing very complex control systems down to the cost of a cheap digital watch, its successors in the middle and late '80s may be large heating appliances that will power a deep freeze and a fridge as a side line. When such equipment is launched on the market, it will no doubt be described as new and revolutionary. Few things ever are.

The 1957 Fridge-Heater.

THE PRESTONS HOUSE IN STAFFORDSHIRE

Mr. & Mrs. Murrey Preston live in Staffordshire in a five-bedroom house built in 1934, where they live with their two children, a cat and the Electricity Council. The children and the cat are permanent fixtures; the Electricity Council arrived in March 1980 and are monitoring the temperatures in different parts of the house and the performance of the Preston's heat pump. This is part of a national study programme involving thirty one ordinary houses with heat pump installations, and the results will probably be published in 1982.

Perhaps one reason why Mr. & Mrs. Preston were approached to help in this way is because at the time they were not particularly energy conservation conscious. They readily admit that they like a home that is really warm, and are not particularly interested in turning down thermostats or turning off radiators in unoccupied rooms. Ten years ago they used 2½ tanks of oil a year to have the house as comfortable as they wished, and the cost of this fuel was a fairly minor element in their budget. Then, like all those who live in large houses with oil fired heating, they suddenly found their heating costs were getting out of hand.

To cut down heat losses they had the walls cavity filled by Rentokil in 1972, and double glazing was fitted all round in 1977. This extra insulation made some improvement, particularly in comfort, but oil prices continued to rise. In Murrey Preston's words, they started to make themselves miserable by niggling about saving oil by playing with the thermostat and turning off radiators. Then he read a magazine article about the use of heat pumps in Canada, and phoned the National Heating Consultancy, asking them who made heat pumps in this country. The N.H.C. advised that Eastwoods in Nottinghamshire were the nearest to the Prestons home. As he has a manufacturing business of his own, Mr. Preston decided that he wanted to see the equipment actually being made, and invited himself over to the factory to have a careful look. He took with him two friends with engineering backgrounds to advise him. The outcome was that he placed a forward order of one of the new Eastwood 14/11 models which were then under development, and had it installed in February 1980. One of his friends is installing a unit in 1981, and the other, who is building a new house, is

The Heating requirement of the Prestons house is 14 kW at 30° F, which is typical for a large pre-war house with cavity insulation. This sort of house is idealy suited to a heat pump.

The old oil boiler, now disconnected, was in the cellar, but the heat pump compressor was put in the garage to suit the new 3-phase electricity supply.

specifying a heat pump to heat it.

The Prestons' heat pump has been trouble-free, and keeps the whole house at the temperature at which the Prestons like it at a completely acceptable cost. In early 1980 the Electricity Council asked manufacturers to suggest installations that could be monitored as part of a heat pump evaluation study, and were put in touch with Mr. Preston. The result is that they now have special temperature recording devices in different rooms in the house, and special electricity meters have been installed (although the Prestons still have to pay for the electricity). The automatic readings are collected every week, and the Prestons are given their own copies. One week's results are reproduced opposite.

These Electricity Council figures show that for the four winter months of November 1st 1980 to March 1st 1981 the heat pump used 9300 units of electricity, of which one third were at the cheap rate. The cost of this was £326.19, or £2.70 per day. The oil boiler did not run at all as the heat pump coped well whatever the weather. For a five bedroomed house built in the '30s this is a very acceptable level of heating cost, and the saving over the earlier oil heating costs is considerable. No direct comparisons can be made, as, like many others who are in business, the Prestons shopped around for oil and bought from the supplier who had the best special offer when they wanted the tank filled. As a result they have no detailed record of oil costs. In eleven months since the machine was installed it has used 15111 units of electricity, $\frac{1}{3}$ of them at the cheap rate. Anticipating that 2600 units will be used in March 1981, while this book is printing, their total for the year will be 17711, providing 46048 kWh of heat. 1476 gallons of oil would have to be burned to provide the same heat, at a cost of £1033 at 70p per gallon. Actual electricity cost is £619, and the saving £414, or 40%. What really matters is that their 1979 oil costs for partially heating the house were unacceptable, and their 1980 costs for full heating are acceptable. The heat pump is there to heat the Prestons as well as the Electricity Council's recording machines; both seem satisfied with what they experience.

TO BE COMPLETED ON SITE AND RETURNED TO THE ELECTRICITY COUNCIL WITH MAGNETIC FILM

Details to be completed by Board's Representative when loading and unloading the Magnetic Film

Consumer's Name ......J. M. PRESTON..Area Board: ..M.E.B..

Address: ....SEEFIELD................District ...Sixim..

................STONE, STAFFS..Project No. 7 6 4

Two Forms MUST be completed if two Contact Meters used

Tariff Meter Readings
N.B. Only read to one decimal place. Do not include meter multiplier, but write it in box below.

METER No.				1 3 7 2 6
SUPPLY (see below)				L O W
FINISH		7 9 6 2 •
START		7 0 5 7 •
ADVANCE		9 0 5 •

METER No.				1 3 7 2 6
SUPPLY (see below)				N O R M A L
FINISH		2 2 1 7 •
START		2 0 3 6 3 •
ADVANCE		1 8 0 8 •

METER No.				8 8 0 3 8
SUPPLY (see below)		L O W	N O R M A L
FINISH	1 4 1 5 6 8 5 0 •
START	1 3 9 2 6 7 6 7 •
ADVANCE	2 3 8 3 •

METER CONSTANT [| | | | •]

SUPPLY NAME
Write in: Total, Unrestricted, Low, Normal, Off-Peak, Day/Night, etc. as appropriate.

Please write any comments overleaf

After replacing the fuse please tick which neons are lit.

	O	1	✗
Neon Panel	✗	2	O
	O	4	O
	O	5	O

LOAD RESEARCH SECTION C32/11/4
THE ELECTRICITY COUNCIL

Contact Meter Readings
Check lead to data recorder and tick box indicating if CM is connected to

METER 1 terminal		or		METER 2 terminal		
DRICO No.				★ 3 2 5 4
MAG FILM N			★ 2 0 0 0 9
TRACK No.				★
CONTACT METER No.	★ 0 3 1 5 3
SUPPLY (see below)				+s
FINISH			8 4 7 • 5
START			7 4 4 • 4
ADVANCE	★	1 0 3 • 1

PLEASE GIVE TIME FROM WATCH
DAY OF WEEK
DATE OF RECORDING

	WATCH TIME	DAY	DATE		
START	1 0 1 0	★ TUE ★	3 2 81
FINISH	1 0 0 5	★ Mon ★	2 3 81
e.g.	13.32 hrs	★ Th ★	11 02 77

DRICO CLOCK TIME	
START TIME	1 0 1 0
FINISH TIME	1 0 1 2

Has the Supply or Installation been altered since last film change.

YES   NO ✓

Signature ...Mtleston...

FOR ELECTRICITY COUNCIL USE ONLY				
CONSUMER No.		★		
MULTIPLIER	★		•	
BATCH OR REEL No				
TEST No				

Low tariff units used in heat pump. Cost £15.29.

Heat pump consumption. Multiplier of 20 gives total 2060 units.

Full rate units for whole houses. 1155 used in heat pump. Cost £50.82. Balance of 653 units used for cooking, lighting, kitchen equipment, etc.

Total cost of running heat pump, February gives £2.36 a day.

Consolidated data for the performance of the installation in February 1981.

Big brother is taking your temperature. One of the Electricity Councils recording thermometers.

The absorber unit of the Prestons heat pump behind the garage.

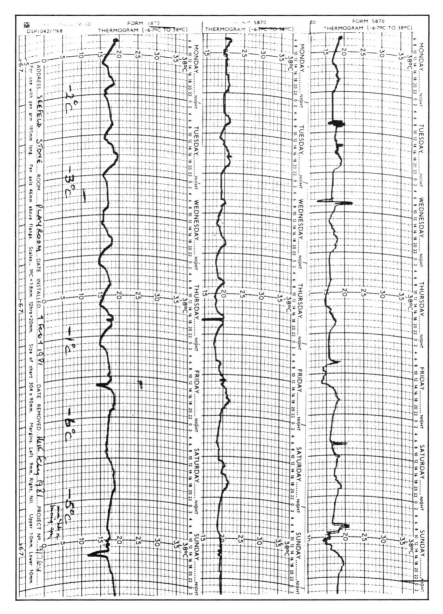

One of the recordings made by the machine shown opposite showing (in centigrade) the steady temperature maintained during the cold week ending February 1981.

A 1980 WARM AIR SYSTEM

This installation in an architect's own home in rural Gloucestershire is a classic example of the use of a heat pump in the right situation. The house is large, extraordinarily well insulated, and lived in by a family who require their heating installation to maintain a controlled temperature in the house, rather than as a source of instant heat available at the flick of a switch.

Michael & Patricia Hall built their large timber-framed house in 1964, building it out from a sloping site with one end supported on stilts. It has an exceptionally high standard of insulation, with four inches of fibre glass and two inches of polystyrene in the walls, ten inches of vermiculite over the ceilings and double glazing throughout. 120 sq.m. of the floor area is designed to be centrally heated, with a rustic family room extension heated only by a log stove. There are high ceilings, and the decor and feel of the house is uncompromisingly modern even today: when newly completed in 1964 it must have been a very avant garde home for the depths of the West Country, and no doubt much admired.

Part of the living area of the Halls modern house in Gloucestershire.

The original central heating was provided by a Lennox oil-fired warm-air system, with a standard arrangement of ducts leading to outlets under the windows, and return air grilles above the doors. The Hall family are outdoor people who do not want particularly high indoor temperatures, and they had hoped that their very well insulated home would always prove economical to heat. By 1979 this was not proving the case, as rising oil prices had lifted their heating costs to a level which Michael Hall felt that he did not want to accept, particularly as he is an architect with a special interest in energy conservation. In an attempt to reduce their energy requirements the Halls turned down the thermostat, and set back the time-clock, and quickly persuaded themselves that some another solution had to be found. The crunch came when the oil-burner started to show signs of its years, and a decision was made to replace it with the current Lennox model of the same type, to be linked with a heat pump that would have its heating coils inside the oil-heater air chamber. The heat pump would be the primary source of warmth, with the oil-burner as a back-up.

This work was carried out in 1980, and the electricity bills which have been carefully kept enable a comparison to be made between the cost of unsatisfactory heating during the 1979/80 winter, and completely satisfactory heating over the same period in 1980/81

The house has a three-phase electricity supply on an Economy 7 tariff with the cheap night rate. This heats the domestic hot water, using an immersion heater controlled with a time switch in a tank large enough to hold a full days hot water requirements.

The heat pump is left on continuously, and is controlled by a room thermostat. This is normally at 60°F. or higher, but is set back manually to 55°F. each night. In the morning it is turned up again, when the oil heater cuts in briefly to meet the sudden heat demand. This is the only time the oil heater ever fires, and Mr. Hall feels that this is a useful arrangement as the response of the warm air system encourages getting up besides exercising the oil boiler and demonstrating that it is in good order. Oil consumption is only 36 gallons a years, although the Halls have yet to experience an exceptionally severe winter, when the oil consumption would be increased. 30 gallons of oil is allowed for in our costs for the six months winter heating period.

The figures are as follows, with all energy consumed costed at March 1981 prices to give proper comparisons.

SIX MONTHS PERIOD, SEPTEMBER 79 — FEBRUARY 1980

360 gallons of oil = 1637 litres at 17p per litre	£278
1728 full rate units of electricity 3145 cheap rate units of electricity, together costing £106 in 1979/80. At March 1981 prices these units would be charged at	£129
TOTAL OIL & ELECTRICITY TO MAINTAIN AN UNSATISFACTORY TEMPERATURE OVER THE 6 MONTHS PERIOD	**£407**

SIX MONTHS PERIOD, SEPTEMBER 80 — FEBRUARY 81

30 gallons of oil = 136 litres at 17p	£23
3561 full rate units 4488 cheap rate units, together costing £232.48	£232
TOTAL OIL & ELECTRICITY TO MAINTAIN A SATISFACTORY TEMPERATURE OVER THE 6 MONTHS PERIOD	**£255**

The 1980/81 costs were 55% of the previous year's costs, the actual cash saving being £152.

An interesting check on the figures is provided by deducting the 1979/80 electricity consumption figures from the 1980/81 figures. This gives an additional consumption, seemingly attributable to the heat pump, of 1836 daytime units and 1343 night units. This is obviously not the whole story, as in spite of higher heat losses from the building at night, the longer daytime tariff period and the higher temperature required during the day would suggest that the daytime consumption should be twice the night-time figure.

The Lennox fuelmaster plus heat pump combined with the standard oil heater from the same manufacturer.

This can be looked at another way. A saving of 330 gallons of oil can be expressed as a saving of 9900 kWh of useful heat, using a round figure conversion factor of 30 kWh per gallon. Taking a coefficient of performance of 2.4 for the Lennox heat pump, the units consumed on both day and night rates is 7630 kWh of useful heat, 23% less than we would expect. There must be an explanation of this — the difference is 2270 kWh of useful heat, which divided by 2.4 gives 946 kWh of electricity — presumably day rate tariff electricity. Here are our missing daytime units from the previous paragraph. If one allocates them to the heat pump, the consumption is 2782 units of daytime electricity, and 1343 units at night - convincingly near to the proportion of 16½ hours to 7½ hours!

The Lennox unit is a Fuelmaster Plus and has been available through Lennox agents for some years. In the Hall's system the actual pump is in an attic with access off a flat roof: their house has a complex roof profile, and the flat roof area is not generally visible from the ground.

A heat store would not be a practical proposition in this installation, for the Lennox unit transfers heat directly from the hot refrigerant coil to the air in the warm air unit, and to introduce water would involve additional heat exchanges, and unacceptable expense. However, the great advantage of a dry system of this sort is that it can be used in reverse to provide air conditioning in very hot weather. This is rarely a major consideration in Britain, particularly for large houses in the country. (Ninety-nine per cent of air conditioning is found in flats in large cities.) Nevertheless it is an added bonus that would be very attractive to anyone with asthma or other health problems, or simply to someone wanting an air-conditioned house!

In considering this installation, it must be kept in mind that the house has an exceptionally high level of insulation, most of which was built in when it was constructed, with some added since. It would not normally be practicable to provide additional insulation to this standard in an existing building. The Hall's saving of 45% in the cost of heating would, if applied to heating costs in a less well-insulated home, result in far larger savings in cash terms.

The Halls heat absorber on the roof.

Architects: Bates, Hall and Partners, 48 Silver Street, Dursley, Glos.

Heating Design Engineer: A.D. Jeffery, 6 Hamble Close, Basingstoke, Hants.

Installation: Heat-Cool Ltd., 110 Wells Road, Bath.

Equipment: Lennox Industries, Lister Road, Basingstoke, Hants.

MYSON HOT WATER GENERATOR

The Myson hot water generator is an extremely compact "package" heat pump designed to be fixed to an outside wall as shown in the photograph. It contains two water circulating pumps, a frost protection thermostat and other features which are essential to a wet heating system, but which are not usually contained in the heat pump casing itself. There are two models, and the smaller is installed in the four-bedroom traditionally built house shown in the photograph. Here it is connected into existing pipework to provide all the heat required in the hot water cylinder and to meet part of the central heating requirement by working in conjunction with a gas-fired boiler.

Mr. & Mrs. J.R. Stocks have lived in the house in Surrey for twelve years, and until 1978 had an oil-fired boiler. This was removed and replaced with the combined gas/heat pump arrangement, and the new installation was completed in two stages. In 1979 when the gas boiler was brought into use, the domestic hot water was heated with an immersion heater on full tariff electricity. The heat pump was installed to take over heating the hot water and to work in harness with the central heating gas boiler in 1980.

Mr. Stocks is a retired engineer who has taken a close interest in the comparative costs of his various heating arrangements, and has kept detailed records of his gas and electricity consumption since the oil boiler was taken out. As he heated his domestic hot water by electricity in 1979, he has been able to calculate his heating requirements for this purpose, and from this to determine the performance of the heat pump in its new role. His figures show that the heat pump reduced the domestic hot water costs by 63%, from £141 to £51. The gas bill has dropped by 11% for the period, and as the family cook with gas, the saving in hours run by the gas boiler must be at least 15%. The electricity bill has of course increased, but off-setting the gas saving against the increased electricity charges, and allowing for inflation, Mr. Stocks calculates that he will have saved £1449 by 1988, which will balance the cost of the heat pump and its installation. This was £1402. These are not very large savings, but they need looking at in detail. The electricity used is on full tariff, and the cost comparisons are against gas costs. Gas is by far the cheapest traditional central heating fuel, and the savings would be even greater if an Economy 7 tariff was adopted. Working through Mr. Stocks calculations again, and computing the savings against the cost of oil, his combined gas/heat pump installation was an excellent investment, and the saving against the cost of oil heating will pay the installation costs by 1988 without allowing for inflation at all.

Equally important is the fact that the householders are convinced that they have got their heating costs under control, and are enthusiastic over the way in which they have achieved this. As a retired couple they anticipate that they will use less of the house on a normal day to day basis in the future, and that the heat pump will thus meet a higher proportion of the cost of heating the smaller living area. They will still have the gas boiler available for occasions when they want to heat the whole house.

The Stocks house in Surrey.

The Myson Hot Water Generator. It can also just be seen in the left of the photograph opposite.

Heating Installation by Gas & Oil Combustion
506/510 Old Kent Road, London SE1.

Hot Water Generator by Myson Copperad Ltd.,
Old Wolverton, Milton Keynes, Bucks MK12 5PT.

18th CENTURY FARMHOUSE IN LEICESTERSHIRE

Geoff White is a control and systems engineer who bought a 200 year old farmhouse on a Leicestershire hilltop in 1979 in a condition which estate agents describe as ''ripe for restoration''. It had three storeys, five bedrooms, and among its many defects there was no central heating of any sort, so that the choice of a system was unhampered. There is no gas available, and Mr. White's day to day work in control engineering enabled him to look at the alternatives with a professional eye.

He was acting as his own architect, planning all the restoration work himself, and doing much of it with his own hands. After a careful study of all the alternatives he decided to install an air-to-water heat pump with a hot water heat store to provide full central heating to 2600 sq.ft. of his home, leaving one wing unheated. The house has solid brick walls which he lined with 25mm high density polyurethane foam and plasterboard, and the top floor ceilings were treated in the same way. This reduced the total heating requirement to 8 kW, which was met by ten radiators carefully sized to suit a heat pump. All the radiators were fitted with thermostatic valves. A 50 gallon domestic hot water tank was installed.

After looking carefully at the alternatives the heat pump chosen was an Eastwood 11/8, rated 8 kW at a 0° external temperature. Geoff White opted for the split model with the compressor in the cellar where he intended to install his heat store, with the absorber unit outside in the kitchen garden.

A major constraint in planning a heat store was a very narrow access to the cellar, which was the obvious place to put it. With a great deal of other work going on at the same time, Mr. White decided that he would not involve himself in fabricating a tank in the cellar, and purchased six 50 gallon insulated hot water tanks, and coupled them up together to provide a single 300 gallon store. They are clearly seen in the photograph. These tanks were relatively expensive, and their total capacity is a little low for the size of the heat pump, 350 to 400 gallons being more appropriate.

The heat pump was ordered with three-phase motors, and the three-phase supply was provided free of charge by the East Midlands Electricity Board. The different Electricity Boards seem to have very different approaches to this: in February 1981 the Yorkshire Electricity Board charged £290 to provide a three-phase supply from a cable only 5 yards from a house for a heat pump installation in their area, while a few miles south the East Midlands Electricity Board was doing the same thing free of charge.

Geoff White is a controls engineer specialising in system design. As was to be expected, he made his own controls for his heating system, and this attracted the attention of Eastwood Heating Developments, who were the manufacturers of his heat pump. With their encouragement and with orders for similar control systems, his company now market three separate multi-function control boxes for heat pump systems, and will advise on all aspects of their installation. The trade name is Church Hill Systems.

The system was brought into full use in February 1980, and the performance has been very carefully monitored. The Electricity Board have loaned an additional Economy 7 meter which has been wired in to record the units consumed by the heat pump installation alone, and this is read daily and the consumption correlated with weather conditions. These figures do not include the electricity used in the house for non-heating purposes.

A particular control feature giving improved economy was not implemented in the first quarter of the year, and it is anticipated that these figures should improve by 2% or 3% in 1981/2.

The Compressors of Geoff Whites heat pump in his cellar.

POWER CONSUMPTIONS AND COSTS FOR THE YEAR
MARCH 1980 — FEBRUARY 1981

	Day Units	Night Units	Cost
1st quarter Mar-May	976.9 at 3.382p	1760.6 at 1.25p	£55.05
2nd quarter June-Aug	118.2 at 3.89p	455.0 at 1.5p	£11.42
3rd quarter Sept-Nov	507.2 at 4.215p	1513.9 at 1.69p	£46.96
4th quarter Dec-Feb	2136.5 at 4.215p	2662.9 at 1.69p	£135.05
TOTAL	3738.8 (36.9%)	6392.4 (63.1%)	£248.48

HEAT PUMP COST (at February 1981 prices)   **£257**
Total Units consumed   10,163
USEFUL HEAT (at 2.4 average C.O.P.)   24,390 kWh
   of which hot water takes   5,200 kWh
   and heating takes   19,190 kWh
PRICE PER USEFUL KWH   **1.05p**

EQUIVALENT OIL CONSUMPTION (at 31 kWh/gallon)   787 gallons
COST OF EQUIVALENT OIL (at 75p per gallon)   **£590**

The saving is £333, and the running costs can be expressed as 44% of the anticipated cost of doing the same job using oil.

If the same electricity consumption had been on a normal tariff the cost would have been £393, or 66% of doing the same job using oil.

Geoff White and the electricity meters from which he logs the performance of his heat pump. The heat store cylinders are in the background.

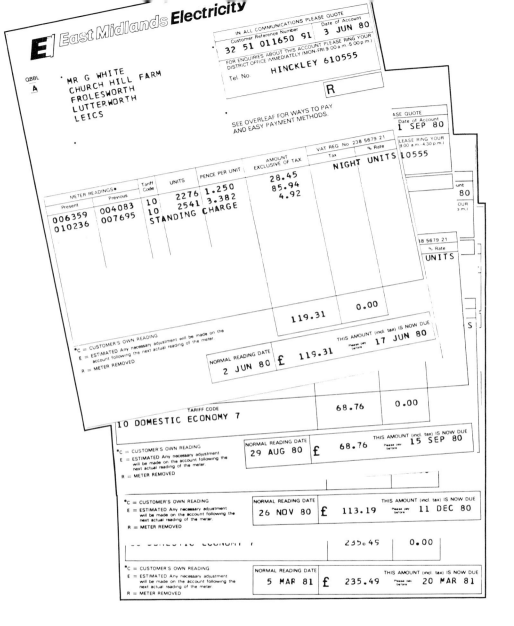

East Midlands **Electricity**

IN ALL COMMUNICATIONS PLEASE QUOTE
Customer Reference Number   Date of Account
32 51 011650 91   3 JUN 80
FOR ENQUIRIES ABOUT THIS ACCOUNT PLEASE RING YOUR
DISTRICT OFFICE IMMEDIATELY (MON-FRI 9.00 a.m.-5.00 p.m.)
Tel. No.   HINCKLEY 610555

QBBL
A

MR G WHITE
CHURCH HILL FARM
FROLESWORTH
LUTTERWORTH
LEICS

R

SEE OVERLEAF FOR WAYS TO PAY
AND EASY PAYMENT METHODS.

ASE QUOTE
Date of Account
1 SEP 80

LEASE RING YOUR
(9.00 a.m.-4.30 p.m.)
610555

VAT REG No 238 5679 21

NIGHT UNITS

METER READINGS•		Tariff Code	UNITS	PENCE PER UNIT	AMOUNT EXCLUSIVE OF TAX	Tax	% Rate
Present	Previous				28.45		
006359	004083	10	2276	1.250	85.94		
010236	007695	10	2541	3.382	4.92		
		STANDING CHARGE					

119.31   0.00

*C = CUSTOMER'S OWN READING
E = ESTIMATED Any necessary adjustment will be made on the account following the next actual reading of the meter.
R = METER REMOVED

THIS AMOUNT (incl. tax) IS NOW DUE
Please pay before 17 JUN 80

NORMAL READING DATE
2 JUN 80 £ 119.31

TARIFF CODE
10 DOMESTIC ECONOMY 7

68.76   0.00

*C = CUSTOMER'S OWN READING
E = ESTIMATED Any necessary adjustment will be made on the account following the next actual reading of the meter.
R = METER REMOVED

NORMAL READING DATE
29 AUG 80 £ 68.76

THIS AMOUNT (incl. tax) IS NOW DUE
Please pay before 15 SEP 80

*C = CUSTOMER'S OWN READING
E = ESTIMATED Any necessary adjustment will be made on the account following the next actual reading of the meter.
R = METER REMOVED

NORMAL READING DATE
26 NOV 80 £ 113.19

THIS AMOUNT (incl. tax) IS NOW DUE
Please pay before 11 DEC 80

235.45   0.00

*C = CUSTOMER'S OWN READING
E = ESTIMATED Any necessary adjustment will be made on the account following the next actual reading of the meter.
R = METER REMOVED

NORMAL READING DATE
5 MAR 81 £ 235.49

THIS AMOUNT (incl. tax) IS NOW DUE
Please pay before 20 MAR 81

Geoff White's Electricity Bills, March 1st 1980 — February 28th 1981
Total of above bills 536.75
Attributable to lighting, cooking, freezer, etc. 279.75
Attributable to heat pump 257.00

During the year the heat system has worked without any problems, providing all the heat required in this large family home. The heat store typically exhausts itself around tea-time, so that the evening heating is on full tariff: if both the heat pump and the heat store had been just a little larger a higher proportion of the electricity used would have been on the night rate, with improved economy and a price per kWh of heat of under 1p. Eastwoods now have a larger pump, designated the 14/11, which is marginally more expensive than the 11/8, which has been installed. Geoff White would have preferred the larger model if it had been available at the time. However, the unit that he has got is giving him exceptionally cheap heating in the type of house that is usually notorious for costing a fortune to heat. His decision in 1979 to invest in a high standard of insulation and his sophisticated heat pump system has given him an average heating cost of £5 per week throughout the year.

Heat Pump by Eastwood Heating Developments
Portland Road,
Shirebrook,
Mansfield, Notts.

Controls by Church Hill Systems
Frolesworth,
Lutterworth, Leics.

A HEAT PUMP DISASTER

Those who have kindly let me feature their heat pumps in this book are generally enthusiastic about the convenience and economy of the installation, and I am grateful to them. I am doubly grateful to someone who has kindly given me data on a heat pump disaster.

In 1973 a builder in Hampshire put up a large house for his own family. With an unusually early understanding of what was going to happen to heating costs he decided to arrange for what was then a very high level of insulation and a heat pump. It is a large house, with a floor area of over 2,500 sq.ft. but the insulation which was provided gave a heating requirement of only 17.1kW - designated in those days as 85,000 B.T.U.

A warm air system was designed with a total of 22 outlets to give temperatures of 75° in the lounge, 70° in three other reception rooms and the hall, and 65° in the five bedrooms. The only domestic heat pump available at that time was an American warm air unit, available in Britain through the European agents. It had various features to suit American conditions, and in particular a supplementary heating arrangement to augment the heat pump output with direct resistance heaters, like electric fires, in the air ducting. This supplementary heating came on automatically whenever the controls sensed that the heat pump was having difficulty in coping with the demand made on it, and had a load of 18.4kW compared with a maximum load of only 7kW for the heat pump itself. The effect of this was to ensure that if the auxiliary heating was ever used to any significant extent, the heating costs for the house would not be the cheapest in Hampshire, but the dearest.

The installation was very carefully monitored by the Electricity Councils research centre at Capenhurst for two full years, and a detailed report was prepared. At that time European interest in heat pumps was centred in Germany, and the report prepared was published in Germany by the Elektrowärm Inst. in 1977. It did not appear in Britain until 1979, when it was summarised in *"Heat Pumps"* by Dr. R.D. Heap, who had conducted the investigations.

The electricity councils automatic recording equipment monitored the electricity consumption and both outside and inside temperatures every half hour for two years. The mass of data obtained was carefully analysed and the conclusion of a lengthy technical report was "the heat pump provided economical and effective heating in this installation, and the owner of the house was well pleased with it". However, deep in the report were the seeds of future problems, and both were clearly identified and described.

The first concern the defrosting arrangements. Winter conditions in the USA are much dryer than those in this country, and the defrosting arrangements in this American machine were designed to suit American conditions. The new house was built in wet wooded country, and the defrosting arrangement had to work overtime. The machine automatically defrosted itself 1379 times in one year. This was not the way in which this part of the equipment was supposed to work.

The second potential problem lay with the energy hungry auxiliary heating arrangements. The Electricity Councils data demonstrated that in the coldest week of the two year study period, from January 28th to February 3rd 1976, the heat pump ran continuously and there was no use of the supplementary heaters. However, in milder weather, when the machine was not running continuously, the supplementary heating was used every time that the machine started up from cold. **The highest electricity demand was in the mildest weather.** In other words, when the heat pump ran continuously it was cost effective because it made no use of the supplementary heaters. Whenever it stopped and was re-started, the supplementary heaters, requiring two and a half times as much electricity, were switched on with it until the house was fully up to temperature.

I called at this house in February 1981, and by chance this was just as the heat pump was being disconnected. The property had changed hands twice, and the new owners had found that their heating bills had been quite unacceptable, totalling £1,462 in the previous twelve months. This was in spite of economising on heating by turning the system off whenever possible. Now they had no possible reason to suspect that by turning off the heating system they were increasing their heating costs, but this is exactly what was happening. In addition to this, a mechanical failure in the heat pump itself had impared its performance, and put even more of the heating load on the auxiliary heaters. The mechanical failure is likely to have been linked with the furious automatic defrosting.

According to the electricity council the heat pump system, including the fans which circulated the warm air in the ducting, required 15084 units to provide a satisfactory level of heating in 1975. 40% of these units were at the cheap rate, as the system ran 24 hours a day. The same units at 1980 prices would have cost £494 - nearly £1,000 less than the actual electricity bills received in 1980.

This cautionary tale emphasises that a householder must be able to understand a heat pump installation if there is one. There is no way that the family concerned with this installation could have known of its peculiarities without having them explained to them. They were not explained to them. Their situation was probably unique, as British made equipment on the market today suits our humid climate, and a modern control system would be designed to prevent auxiliary heaters meeting a start up demand.

One practical point - if you are having a heat pump installed with auxiliary heaters, ask for a warning light to be provided to signal when the auxiliary heaters have switched themselves on, and have this light somewhere where it can be seen - perhaps in the kitchen. If an oil fired boiler is leaking oil you can see it, and a leak of gas from a gas heater becomes obvious one way or another, but wasted electricity can easily be undetected. A signal light will enable you to keep a check.

The house in Hampshire in which the heat pump was installed in 1973.

A HEAT PUMP IN THE ROOF

Eric Tutton is a retired engineer who lives at Harlech in West Wales and has kept full records of the performance of the heat pump which was installed in his bungalow in 1979. His home, seen in the photograph below, is in an exposed situation, and the average oil consumption to heat it over the last three years was 807 gallons a year, costing £562 per annum at 1980 prices. As a professional engineer he was aware of the possibilities of heat pumps: after a working life spent as a manager in the confectionery industry their economics appealed to him: and as an enthusiastic model engineer in his retirement the idea of monitoring the performance of a heat pump system had an immediate attraction. His oil boiler was taken out and sold, and its place in the kitchen taken by the compressor unit of an Eastwood Model 11/8. This is a standard air-to-water unit, and Mr. & Mrs. Tutton had it connected to their existing radiator system and have no other central heating arrangements. Where they did something different was where they put the absorber unit.

The Tutton's bungalow at Harlech.

The bungalow has a slate roof which absorbs a great deal of solar radiation. The eaves are not vented (as would be the case in a building constructed to current Building Regulations), so Mr. Tutton decided to put his absorber unit in the loft and to build a timber duct to take the cold air leaving the unit right outside, so that incoming air to replace it would have to find its way in between the slates. He hoped that these would act as several hundred miniature solar panels and warm the incoming air. Although installers do not usually recommend putting equipment in roof spaces for a variety of practical reasons, — of which ease of access is one — the whole thing works splendidly.

Daily records have been kept of electricity consumed, external temperatures, and the temperature in the roof. Two temperature readings are taken a day, at 7.30 a.m. and at 12.30 p.m., and the figures given are the mean of all the months readings. Electricity is on an Economy 7 tariff, and 40% of the consumption is at the cheap rate.

The room thermostat is set at 65° F. and the pump setting is 150° F. The system is controlled by a time clock that turns it on at 4 a.m. and off at 10 p.m., exactly the same hours that the oil fired system used to work. If the heat pump is unable to maintain a hot water cylinder temperature of 140° F, a 3kW immersion heater is allowed to cut in in the early morning while the system is still on the night rate tariff.

The heat pump compressor in the kitchen taking the place of the oil boiler. Eric Tutton now uses the boiler flue pipe to extract air from the kitchen to feed to the heat absorber in the roof.

Mr. Tutton's figures are reproduced opposite, and it is interesting to see that his roof space is consistently warmer than the outside air even in winter in spite of the continuous movement of air via the heat pump. This is clear evidence of the solar gain from the slates (which depends on available radiation, not necessarily on direct sunshine). The manufacturer's performance graph for the Eastwood 14/11 show that his extra 6°F. in his roof in December gave an extra 1kW of output — free of charge.

The absorber in the roof is suspended and does not rest on the roof timbers in order to avoid any possibility of vibration or noise. In practice the whole installation is in no way obtrusive, and the Tuttons consider the noise level to be the same as that of the old oil boiler.

The heat pump was purchased for just under £1000 in 1979, and Mr. Tutton spent £180 on materials for the ducting. With a saving at current fuel costs of £208 per annum, the system will have paid for itself in less than six years, even without any allowance for inflation.

The heat absorber in the roof.

ERIC TUTTON'S HEAT PUMP RECORDS

SYNOPSIS

Month	Units used (Heat pump only)	Mean outside Temperature °F.	Roof space Temperature °F.
April 80	773	49.1	60.4
May 80	505	57.4	69.4
June 80	381	56.7	71.1
July 80	393	56.1	68.0
Aug 80	362	59.9	69.1
Sept 80	373	55.2	62.4
Oct 80	786	48.4	55.9
Nov 80	1117	44.6	51.6
Dec 80	1316	43.2	48.0
Jan 81	1315	43.3	47.1
Feb 81	1224	39.6	46.9
Mar 81	1112	45.9	53.9

TOTAL UNITS USED 9657

Average oil consumption 1976-77-78	803 galls p.a.
Cost of 803 gall at 70p	£562
Actual cost of 9657 units, 40% at cheap rate	£326
Saving	**£236 (42%)**

HEATING A SWIMMING POOL WITH A CALOREX POOL HEATER

The swimming pool illustrated is at the Essex home of Mr. & Mrs. Michael Patten. Even though the photograph was taken in winter, it makes it clear that this is the dream pool visualised by most people who have large gardens and a mistaken idea that building a swimming pool to reduce the area of lawn to be cut is a labour saving investment!

The pool is 38' x 16' with a 10' x 5' Roman bay, and has a capacity of 20,000 gallons. It is used continuously from May through to the end of September each year, and a minimum temperature of 81°F. is maintained through most of the season, with temperatures sometimes rising to 92°F. A floating bubble cover is used to minimise heat losses.

An oil heater and then two successive heat pumps have been installed. The first heat pump was a Calorex Model 750, and was installed in 1978. It did all that an earlier oil-fired installation had done with such a saving in running costs that in 1980 the Pattens decided to use their trade connections to sell it second-hand to another pool owner who had admired it, and to buy one of the new Calorex 800 Models which had just come on the market. This is left switched on all the time with the thermostat set at 81° F, and it only has to run for six or seven hours each day once the pool is up to temperature.

When the ambient temperature is 55° F. the output of the Calorex 800 is 9.6 kW against an electricity consumption of 2.4 kW. At 65° F. this rises to 13.0 kW output for 2.65 kW of electricity. The annual cost over the five months heating season is £250, and based on previous experience the cost of doing the same job with oil at today's prices would be around £625.

The heat pump, situated at the side of the old boiler room, gives a clear idea of the difference in the size of the two heating systems.

The electricity used is on the normal domestic tariff: if an Economy 7 tariff was available the savings would be even larger.

When discussing his pool heating with the author, Mr. Patten particularly emphasised the importance that he places on the heat pump being maintenance free and saving the bother of oil deliveries and the expense of routine boiler maintenance.

This is another example of a heat pump which not only makes cash savings in real terms, but which also enables a facility to be enjoyed in a way which would not be possible if the owners had to contend with oil prices.

Heat Pump by Calorex Heat Pumps,
Unit 2,
The Causeway,
Malden,
Essex CM9 7PU.

APPENDICES

HEAT PUMPS — THE THEORY & THE JARGON

This is not an essay on the principles of refrigeration engineering but simply an explanation of some of the concepts which the reader was asked to take for granted in Chapter One. Against the advice of my engineer friends, I persist in using Fahrenheit temperatures: those who can relate to Centigrade can make the conversions and probably know everything in this Appendix anyway.

To keep things simple this appendix deals with the operation of heat pumps that collect heat from the air, and use it to provide hot water for a radiator and a hot water cylinder. They work by reducing the temperature of the outside air by a few degrees, and using the heat taken from it to heat the water which is pumped around the radiators.

This depends on the principle that when a liquid evaporates it absorbs heat from its surroundings. This heat is stored in the vapour until the vapour condenses, when it gives up this heat.

A heat pump has a closed circuit of pipes containing a special chemical which is liquid when compressed but which vapourises when it passes through a nozzle into a low pressure part of the circuit. This chemical is described as a refrigerant. There is a compressor in the system which keeps the refrigerant moving around. After it has been compressed, the pipes are surrounded by water which collects the heat given up, and at the point where it is allowed to vapourise, the pipes are surrounded by the current of air from which they absorb heat.

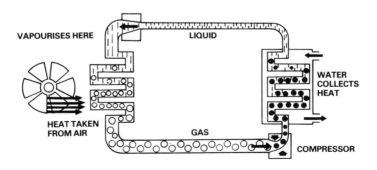

The process is shown diagramatically below. This is so simplified that it can be likened to explaining Concorde by making a paper dart, but it serves. The essential thing to visualise is that the refrigerant is circulating all the time, that it is a vapour when it enters the compressor, that when compressed it gets very hot and gives up its heat to a water jacket, condensing in the process, and the liquid refrigerant finally escapes through a narrow jet into a low pressure part of the system. It vapourises on leaving the jet, and gets very cold in doing so. The cold vapour then picks up heat from air which is blown round the pipes in which it is contained, and keeps moving until it reaches the compressor where the cycle starts again.

The only energy used in the process is the electricity used to drive the compressor and the fan. The efficiency of a heat pump is commonly measured by its coefficient of performance (COP) which is the ratio of energy input to heat output. This will depend on both the ambient temperature and on the temperature of the hot water delivered by the machine.

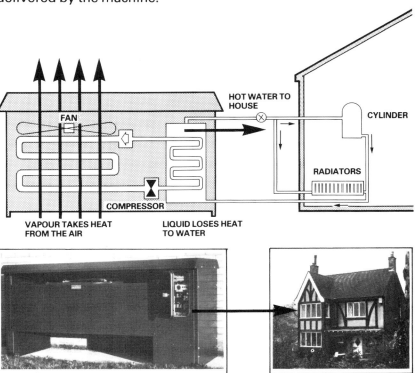

This is shown in the top graph opposite, where the COP of a typical modern machine is plotted against the ambient air temperature. Two curves are shown, one for a 104°F. water delivery, and the other for a 140°F. water delivery.

Let us apply this to our domestic heating requirements. In an ordinary year our houses are heated for approximately thirty-three weeks, and our average family wants its central heating to operate when the ambient temperature falls below 61°F. The average winter temperature is 43°F. and there are normally ten days when the average temperature is below 30°F. The heating requirement of a house is expressed as the kilowatts required to maintain an internal temperature of 70°F. in living rooms and 60°F. elsewhere at an external temperature of 32°F. The heat input required for the property will vary directly with the ambient temperature, and for a small four-bedroomed detached house with a reasonable standard of insulation, the heat requirements are shown by the line on the second graph.

The solid line shows the heat available as water at 140°F. from a typical modern heat pump — in this case an Eastwood Model 80-14/11. (The last two figures of the manufacturer's code refer to the output at ambient air temperature of 43°F. and 32°F. respectively — i.e. the machine will produce 14 kW at 43°F, and 11 kW at 32°F.) The point where this performance curve crosses the heating requirement line is called the *balance point.* Broadly speaking if this occurs at freezing point or below the heat pump is the right size to heat the house on its own, using a booster system for the few very cold days of each year. If the balance point is above freezing, a bivalent system will be more suitable. In the above example the heat pump will provide full heating down to 32°F, and statistics on our weather indicate that there will still be approximately fourteen days in the year when the required internal temperatures cannot be maintained by the heat pump alone. The short-fall can be made up by using the Eastwoods 3 kW boost heater, or an open fire or electric heater, and switching off radiators in this room to allow the heat pump to heat the remainder of the house. The alternative is a larger heat pump.

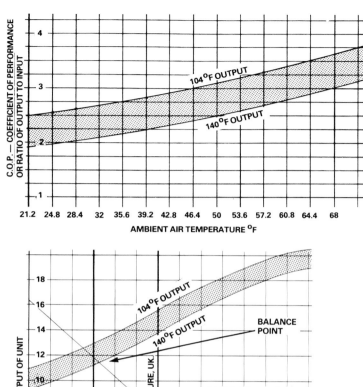

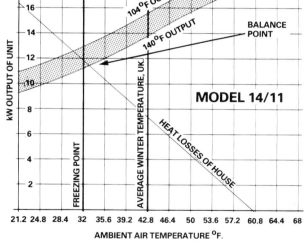

Graphs of this sort, and consideration of balance points, are features of virtually every heat pump manufacturer's brochure. Balance points may also feature in a heating engineer's report. This appendix may help to make them intelligible, but it is a poor substitute for making your heating engineer sit down and explain the whole thing himself.

ELECTRICITY BOARD TARIFFS
as advised at April 1981

All tariffs involve a quarterly standing charge plus a charge per unit.

| | STANDARD TARIFF | | ECONOMY 7 TARIFF | | |
| | Standing | Unit | Standing | Day rate Unit | Night rate Unit |
	£	p	£	p	p
London Electricity Board 46 New Broad Street London EC2M 1LS Tel: 01 588 1280	7.00	5.02	8.82	5.32	1.82
South Eastern Electricity Board Queen's Gardens Hove Sussex BN3 2LS Tel: Brighton 724522 (0273)	6.89	4.50	8.48	4.58	1.82
Southern Electricity Board Southern Electricity House Littlewick Green Nr. Maidenhead Berks SL6 3QB Tel: Littlewick Green 2166 (062 882)	6.24	4.47	8.06	4.81	1.82
South Western Electricity Board Electricity House Colston Avenue Bristol BS1 4TS Tel: Bristol 26062 (0272)	6.45	4.9	8.20	5.26	1.82
Eastern Electricity Board PO Box 40 Wherstead Ipswich Suffolk IP9 2AQ Tel: Ipswich 55841 (0473)	5.85	4.44	7.54	4.74	1.82
East Midlands Electricity Board PO Box 4 North PDO 398 Coppice Road Arnold Nottingham NG5 7HX Tel: Nottingham 269711 (0602)	5.50	4.48	7.25	4.48	1.82
Midlands Electricity Board PO Box 8 Mucklow Hill Halesowen West Midlands B628BP Tel: 021 422 4000	5.30	4.71	7.50	4.85	1.82
South Wales Electricity Board St. Mellons Cardiff CF3 9XW Tel: Cardiff 792111 (0222)	6.15	4.84	7.90	5.15	1.82
Merseyside & North Wales Electricity Board Sealand Road Chester CH1 4LR Tel: Chester 377111 (0244)	5.59	4.89	7.70	5.33	1.82
Yorkshire Electricity Board Scarcroft Leeds LS14 3HS Tel: Leeds 892123 (0532)	5.45	4.70	7.05	5.00	1.82

North Eastern Electricity Board Carliol House Newcastle upon Tyne NE99 1SE Tel: Newcastle 27520 (0632)	5.35	4.85	7.10	5.15	1.82
North Western Electricity Board Cheetwood Road Manchester M8 8BA Tel: 061 834 8161	5.10	4.68	6.84	5.01	1.82
South of Scotland **Electricity Board** Cathcart House Inverlair Avenue Glasgow G44 4BE Tel: 041 637 7177	Min. charge	4.13	Min. charge	4.41	1.85
North of Scotland **Hydro-Electric Board** 16 Rothesay Terrace Edinburgh EH3 7SE. Tel: 031 225 1361	4.35	Varies	9.75	4.14	1.76
Northern Ireland **Electricity Service** 120 Malone Road Belfast BT9 5HT Tel: Belfast 661100 (0232)	6.60	5.1	0.60	5.1	2.45

These tariffs are to come into force on April 1st 1981. Most of the costs quoted in this book are based on the old Economy 7 tariff of 1.69p, as the new charges were not announced until the main part of the book had been printed.

ADDRESSES FOR FURTHER INFORMATION

MANUFACTURERS

Manufacturers of British made heat pumps will arrange for accredited installers or their own representatives to present proposals for the use of their equipment in domestic premises. Leading manufacturers are, in alphabetical order.

Calorex Heat Pumps
Unit 2
The Causeway
Malden
Essex   (0621-56611)   Heat pumps for swimming pools.

Church Hill Systems
Frolesworth
Lutterworth   Special control systems for heat
Leics.   (0455-202314)   pumps.

Eastwood Heat Pumps
Portland Road
Shirebrook
Mansfield   Heat pumps for all domestic
Notts.   (0623-853221)   applications.

Glynwed Ltd
Cranmore Drive
Shirley
Solihull   Crusader Heat pump for domestic
W. Midlands   (021-704-4322) hot water, not central heating.

Lennox Industries Ltd
Lister Road
Basingstoke   Heat pumps for all domestic
Hants.   (0256-61261)   applications.

Myson-Copperad Ltd
Old Wolverton
Milton Keynes   Heat pumps for all domestic
Bucks.   (0908-312641)   applications.

IMPORTED HEAT PUMPS

There are many agents for foreign heat pumps. Leading importers are —

Trace Cleveland Ltd
Unit 6
Industrial Estate West
Witham
Essex   (0376-515511)

Weathermaker Ltd
Churchill House
Talbot Road
Manchester   (061-872-7035)

Conservatherm Limited
Unit 18
Huffwood Trading Estate Ltd
Billinghurst
W. Sussex   (0403-814168)

Electraplant Limited
Glebe Works
Braunston Road
Oakham, Leics.   (0572-3831)

Steibel Elktron Ltd
25 Lyveden Road
Brackmills
Northampton   (0604-66241)

FURTHER INFORMATION FROM:

Heat Pump Information Service
161 Drury Lane
London WC2   (01-242-3706)

Publicity service financed by manufacturers and importers.

The Build-Electric Bureau
26 Store Street
London   (01-580-4986)

Electricity Council demonstration systems which can be inspected.

Heating & Ventilation Contractors Assn.
E.S.C.A. House
32 Palace Court
Bayswater
London   (01-229-2488)

Publish "Guide to Good Practise for Unit Air Conditioning including Heat Pumps" and provide a list of contractors.

Design & Materials Ltd.,
Carlton Industrial Estate,
Worksop, Notts. (0909 730333)

Architectural service, find builders, supply materials for low energy houses & bungalows suitable for Heat Pumps.

APPENDIX FOUR
BIBLIOGRAPHY

"Domestic Heat Pumps", J.A. Sumner, Prism Press, 1976

John Sumner spent thirty years warning of an energy crisis and promoting the heat pump idea. He was involved in many experimental installations in the '50s and '60s. He died in 1980, by which time his vision of the heat pump as an accepted heating appliance had become a reality. His book mixes his philosophy with advanced mathematics in a delightful way, but is not easy reading. Essential for the serious student.

"Heat Pumps", R.D. Heap, Spon, 1979

Dr. Heap wrote his book on retirement from the Electricity Council, where for eleven years he worked on heat pumps at the Council's research centre at Capenhurst. A comprehensive and up to date review of all aspects of heat pump technology.

"Heat Pumps — Design and Application", Reay & MacMitchel, Pergamon, 1979

A 300-page reference book with most comprehensive references to source material. A standard work for researchers and students.

"Heat Pump Design", The Electricity Council, 1978

One of a number of publications by the Council's Environmental Engineering Section, based on their investigations into the efficiency of various installations. Unbiased data, very professionally analysed.

"Heat Pumps — The Energy Savers", The Electricity Council

A 24-page colour booklet produced to encourage the use of heat pumps. Very clear explanation of basic principles.

Department of the Environment — Building Research Establishment Paper CP 19/76, HMSO, 1976

The Government's view. The figures are now five years old.

APPENDIX FIVE

CENTIGRADE TO FAHRENHEIT CONVERSION TABLE

C°	F°	C°	F°	C°	F°
— 6	21.2	18	64.4	42	107.6
— 5	23.0	19	66.2	43	109.4
— 4	24.8	20	68.0	44	111.2
— 3	26.6	21	69.8	45	113.0
— 2	28.4	22	71.6	46	114.8
— 1	30.2	23	73.4	47	116.6
0	32.0	24	75.2	48	118.4
1	33.8	25	77.0	49	120.2
2	35.6	26	78.8	50	122.0
3	37.4	27	80.6	51	123.8
4	39.2	28	82.4	52	125.6
5	41.0	29	84.2	53	127.4
6	42.8	30	86.0	54	129.2
7	44.6	31	87.8	55	131.0
8	46.4	32	89.6	56	132.8
9	48.2	33	91.4	57	134.6
10	50.0	34	93.2	58	136.4
11	51.8	35	95.0	59	138.2
12	53.6	36	96.8	60	140.0
13	55.4	37	98.6	61	141.8
14	57.2	38	100.4	62	143.6
15	59.0	39	102.2	63	145.4
16	60.8	40	104.0	64	147.2
17	62.6	41	105.8	65	149.0

APPENDIX SIX

ACKNOWLEDGEMENTS

The Author wishes to express his appreciation of the help and advice received from many sources in writing this book, and in particular from

Dr. Geoff Brundrett, The Electricity Council

John Churley Esq., Coronet Heat Pumps Ltd.

Bill Eastwood Esq., Eastwood Heating Developments Ltd.

Mr. & Mrs. M.W. Hall

Mike & Barbara Mawer

Mrs. T.J. McGinn

Denis O'Brien Esq., Lennox Industries Ltd.

Michael Patten Esq.

Mrs. K. Plested

Murrey Preston Esq.

Steve Rich Esq.,

Paul Scott Esq., Myson Copperad Ltd.

Eric Tutton Esq.

Keith Walker Esq., Yorkshire Heating Supplies.

Geoff White Esq., Church Hill Systems Ltd.

INDEX